BOB MILLER'S ALGEBRA FOR THE CLUELESS

ALGEBRA

OTHER BOOKS BY BOB MILLER

BOB MILLER'S ALGEBRA FOR THE CLUELESS

ALGEBRA

Robert Miller

Formerly of the Mathematics Department
City University of New York

Second Edition

McGraw-Hill

New York   Chicago   San Francisco   Lisbon   London   Madrid
Mexico City   Milan   New Delhi   San Juan   Seoul
Singapore   Sydney   Toronto

To my dearest Marlene, I dedicate this book and everything else I ever do to you. I love you very much.

The **McGraw·Hill** Companies

Library of Congress Cataloging-in-Publication Data

Miller, Robert, date.
  Bob miller's algebra for the clueless : algebra / Robert Miller.—2nd ed.
     p. cm.—(Bob Miller's clueless series)
  Includes index.
  ISBN 0-07-147366-1 (alk. paper)
  1. Algebra. I. Title.
  QA152.3.M578 2007
  512—dc22

                                          2006008455

1 2 3 4 5 6 7 8 9 0 DOC/DOC 0 1 0 9 8 7 6

ISBN 0-07-147366-1

*The sponsoring editor for this book was Barbara Gilson and the production supervisor
was Pamela A. Pelton. It was set in Melior by North Market Street Graphics. The art
director for the cover was Gary Brumberg.*

Printed and bound by RR Donnelley.

This book was printed on acid-free paper.

McGraw-Hill books are available at special quantity discounts to use as premiums and
sales promotions, or for use in corporate training programs. For more information,
please write to the Director of Special Sales, McGraw-Hill Professional, Two Penn Plaza,
New York, NY 10121-2298. Or contact your local bookstore.

TO THE STUDENT:

This book is written for you: not for your teacher, not for your next-door neighbor, not for anyone but you.

Unfortunately, most math books today teach algebra in a way that does not give you the basics you need to succeed. Many students immediately have problems, while some manage to succeed, only to have problems in algebra 2 or precalculus. This book gives and explains the topics you will need to succeed.

However, as much as I hate to admit it, I am not perfect. If you find something that is unclear or a topic that should be added to the book, you can contact me in one of two ways. You can write me c/o McGraw-Hill, Two Penn Plaza, New York, NY 10121-2298. Please enclose a self-addressed stamped envelope. Be patient; I will answer. You can also see me at www.bobmiller.com and contact me at bobmiller@mathclueless.com. I will answer faster than if you write, but again, please be patient.

If you need more advanced stuff, there is *Geometry for the Clueless, Precalc with Trig for the Clueless,* and *Calc I, Calc II,* and *Calc III for the Clueless.* If you are preparing for the SAT, *SAT® Math for the Clueless* will help you.

Now enjoy the book and learn!!!

Bob Miller

CONTENTS

CHAPTER 1

NATURAL NUMBERS AND INTRODUCTORY TERMS

CONGRATULATIONS

Congratulations!!!! You have reached a point that most of the world does not even come near, believe it or not. You are starting algebra. It is a great adventure we are beginning.

Algebra is a new subject, even if you had a little in the past. You may have some trouble at the beginning. I did too!!!! Even though I was getting almost everything correct, for more than two months I didn't really understand what was happening, really!!!! After that things got better. Next there are new vocabulary words. There are always some at the start of a new course. In algebra there are less than 100. (In English you need about 7000 new words for high school.) Since there are so few words, *every* word is very important. You must not only memorize the words but also understand them. Many of these words occur right at the beginning. This may be kind of boring, but learning these words is super necessary. If you need to review your fractions, decimals, percents, and graphs, look at the appendix at the back of the book.

Now relax. Read the text slowly. If you have trouble with an example, write it out and don't go to the next step until you understand the previous step.

I really love this stuff. I hope after reading parts of this book, you will too.

Okay. Let's get started.

INTRODUCTORY TERMS

At the beginning, we will deal with two sets of numbers. The first is the set of *natural numbers* nn, which are the numbers 1, 2, 3, 4, . . . and the second is the set of *whole numbers* 0, 1, 2, 3, 4, ... The three dots at the end mean the set is *infinite,* that it goes on forever. The first four numbers show the pattern. Numbers like 5.678, 3/4, –7/45, $\sqrt{7}$, π, and so on are not natural numbers and not whole numbers.

We will talk about *equality statements,* such as 4 + 5 = 9 and 7 – 3 = 4.

We will write 3 + 4 ≠ 10, which says 3 plus 4 does not equal 10.

A *prime* natural number is a natural number with two distinct natural number factors, itself and 1. 1 is not a prime. The first eight prime factors are 2, 3, 5, 7, 11, 13, 17, and 19.

9 is not a prime since it has three nn prime factors, 1, 3, and 9. Numbers like 9 are called *composites*.

The *even* natural numbers are the set 2, 4, 6, 8, ...

The *odd* natural numbers are the set 1, 3, 5, 7, ...

We would like to graph numbers. We will do it on a *line graph* or *number line.* Let's give some examples.

EXAMPLE 1—

Graph the first four even natural numbers.

First, draw a straight line with a ruler.

Next, divide the line into convenient lengths.

Next, label 0, called the *origin,* if practical.

Now do the problem.

```
  |   |   •   |   •   |   •   |   •   |
  0   1   2   3   4   5   6   7   8   9
```

EXAMPLE 2—

Graph all the even natural numbers.

```
                              ...
  |   |   •   |   •   |   •   |   •
  0   1   2   3   4   5   6   7   8
```

The three dots mean the set is infinite.

EXAMPLE 3—

Graph all the natural numbers between 50 and 60.

```
  |   •   •   •   •   •   •   •   •   •   |
  50  51  52  53  54  55  56  57  58  59  60
```

The word *between* does *not, not, not* include the end numbers. In this problem, it is not convenient to label the origin.

EXAMPLE 4—

Graph all the primes between 40 and 50.

```
  |   •   |   •   |   |   |   •   |   |   |
  40  41  42  43  44  45  46  47  48  49  50
```

EXAMPLE 5—

Graph all multiples of 10 between 40 and 120 inclusive.

```
  |   •   •   •   •   •   •   •   •   |
  30  40  50  60  70  80  90 100 110 120 130
```

Inclusive means both ends are part of the answer.

nn multiples of 10: take the natural numbers and multiply each by 10.

Because all of these numbers are multiples of 10, we divide the number line into 10s.

A *variable* is a symbol that changes. In the beginning, most letters will stand for variables. Later, letters toward the end of the alphabet will stand for variables.

A *constant* is a symbol that does not change. Examples are 5, 9876, $\sqrt{}$, π, +, . . . are all symbols that don't change. Later, much later, letters like a, b, c, and k will stand for constants, but not now.

We also need words for addition, subtraction, multiplication, and division. Here are some of the most common:

Addition—Sum (the answer in addition), more, more than, increase, increased by, plus

Subtraction—Difference (the answer in subtraction), take away, from, decrease, decreased by, diminish, diminished by, less, less than

Multiplication—Product (the answer in multiplication), double (multiply by 2), triple (multiply by 3), times

Division—Quotient (the answer in division), divided by

Let's give some examples to learn the words better.

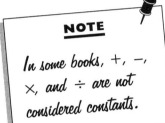

NOTE

In some books, +, −, ×, and ÷ are not considered constants.

EXAMPLE 6—

The sum of f and 6

ANSWER

f + 6   or   6 + f

In addition the order does not matter because of the *commutative law,* which says that the order in which you add does not matter.

a + b = b + a

4 + 3 = 3 + 4

Subtraction is the one that always causes the most problems. Let's see the words.

EXAMPLE 7

The difference between x and y	x − y
x decreased by y	x − y
x diminished by y	x − y
x take away y	x − y
x minus y	x − y
x less y	x − y
x less than y	y − x
x from y	y − x

Verrry important. Notice that "less" does *not* reverse while "less than" reverses. 6 less 2 is 4, while 6 less than 2 is a negative number, as we will see later. (As you read each one, listen to the difference!)

Also notice that subtraction is *not* commutative, because 7 − 3 ≠ 3 − 7.

EXAMPLE 8

The product of 6 and 4

ANSWER

6(4) or (6)(4) or (6)4 or (4)6 or 4(6) or (4)(6)

Verrrry important again. The word *and* does *not* mean addition. Also, see that multiplication is commutative:

ab = ba

(7)(6) = (6)(7)

EXAMPLE 9—

    a. s times r   rs

    b. m times 6   6m

ANSWER

    a. Although either order is correct, we usually write products alphabetically

    b. Although either order is again correct, we always write the number first.

EXAMPLE 10—

6 divided by m

ANSWER

$$\frac{6}{m}$$

For algebraic purposes, it is almost always better to write division as a fraction.

EXAMPLE 11—

The difference between and a and b divided by p

ANSWER

$$\frac{a - b}{p}$$

EXAMPLE 12—

6 less the sum of h and m

ANSWER

$6 - (h + m)$

The ( ) symbols are parentheses, the plural of parenthesis. [ ] are brackets. { } are braces.

There are shorter ways to write the product of identical factors. We will use *exponents* or *powers.*

y^2 means (y)(y) or yy and is read, "y squared" or "y to the second power." The 2 is the exponent or the power.

8^3 means 8(8)(8) and is read, "8 cubed" or "8 to the third power."

x^4 means xxxx and is read "x to the fourth power."

x^n means (x)(x)(x) . . . (x) (n factors) and is read, "x to the nth power."

$x = x^1$, x to the first power.

I'll bet you weren't expecting a reading lesson. There are always new words at the beginning of any new subject. There are not too many later, but there are still some more now. Let's look at them.

$5x^2$ means 5xx and is read, "5x squared."

$7x^2y^3$ is 7xxyyy, and is read, "7x squared y cubed."

$(5x)^3$ is (5x)(5x)(5x) and is read, "the quantity 5x, cubed." It also equals $125x^3$.

EXAMPLE 13

Write in exponential form:

   a. (3)(3)(3)xxxxx

   b. aaabcc

   c. $(x + 6)(x + 6)(x + 6)(x - 3)(x - 3)$

ANSWERS

   a. 3^3x^5

   b. a^3bc^2

   c. $(x + 6)^3(x - 3)^2$

EXAMPLE 14

Write in completely factored form with no exponents:

   a. $28a^4bc^3$

   b. $6(x + 6)^3$

ANSWERS

   a. $(2)(2)(7)aaaabccc$

   b. $(2)(3)(x + 6)(x + 6)(x + 6)$

ORDER OF OPERATIONS, NUMERICAL EVALUATIONS, AND FORMULAS FROM THE PAST

Suppose we have $4 + 3(4)$. This could mean $7(4) = 28$ orrrr $4 + 12 = 16$. Which one? In math, this is a no-no! An expression can have only one meaning. The *order of operations* will tell us what to do first.

1. Do any operations inside parentheses or on the tops and bottoms of fractions.

2. Do exponents.

3. Do multiplication, left to right, as it occurs.

4. Do addition and subtraction.

EXAMPLE 1

Our first example:

$4 + 3(4) = 4 + 12 = 16$

because multiplication comes before addition.

EXAMPLE 2—

$5^2 - 3(5 - 3) + 2^3$ **Inside parentheses first**

$= 5^2 - 3(2) + 2^3$ **Exponents**

$= 25 - 3(2) + 8$ **Multiplication, then adding and subtracting**

$= 25 - 6 + 8 = 27$

EXAMPLE 3—

$32 \div 8 \times 2$ **Multiplication and division, left to right, as they occur. Division is first.**

$= 4 \times 2 = 8$

EXAMPLE 4—

$$\frac{4^3 + 6^2}{12 - 2} + \frac{8(4)}{18 - 2} = \frac{64 + 36}{12 - 2} + \frac{8(4)}{18 - 2} = \frac{100}{10} + \frac{32}{16}$$

$$= 10 + 2 = 12$$

Sometimes we have a step before 1. Sometimes we are given an *algebraic expression,* a collection of factors and mathematical operations. We are given numbers for each variable and asked to *evaluate,* find the numerical answer for the expression. The steps are:

   0. Substitute in parentheses, the value of each letter.

   1. Do inside parentheses and the tops and bottoms of fractions.

   2. Do each exponent.

   3. Do multiplication and division, left to right, as it occurs.

   4. Lastly, do all adding and subtracting.

EXAMPLE 5—

If $x = 3$ and $y = 2$, find the value of:

  a. $y(x + 4) - 1$

  b. $5xy - 7y$

  c. $x^3y - xy^2$

ANSWERS

  a. $y(x + 4) - 1 = (2)(3 + 4) - 1 = (2)(7) - 1 = 14 - 1 = 13$

  b. $5xy - 7y = 5(3)(2) - 7(2) = 30 - 14 = 16$

  c. $x^3y - xy^2 = (3)^3(2) - (3)(2)^2 = (27)(2) - 3(4) = 54 - 12 = 42$

Let's finish this section by reviewing some geometric facts from the past.

Square

$s = $ side

Area $A = s^2$

Perimeter $p = 4s$

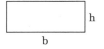

Rectangle

$b = $ base

$h = $ height

$A = bh$

$p = 2b + 2h$

Triangles

Quite a bit on triangles.

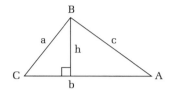

b = base

$$A = \frac{1}{2} \, bh$$

h = height, perpendicular (90° angle) to the base

$$p = a + b + c$$

Sides denoted by small letters

Sum of interior angles = 180°

Angles denoted by capital letters

$$\angle A + \angle B + \angle C = 180°$$

Angle A opposite side a

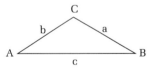

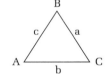

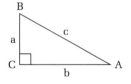

Isosceles triangle:
a triangle with two equal sides
a = b; equal sides are the legs
c = base; could be smaller
  than, equal to, or bigger
  than a leg!
$\angle A = \angle B$ base angles are
  equal
$\angle C$ vertex angle

Equilateral triangle:
a triangle with all
sides equal
a = b = c;
also equiangular
$\angle A = \angle B = \angle C = 60°$

Right triangle:
a triangle with one right
angle
$\angle C$, a right angle = 90°
c = hypotenuse
a, b legs equal or
unequal

Let's recall the names of angles:

Acute angle	Greater than 0° buuuut less than 90°
Right angle	Exxxxactly 90°
Obtuse angle	More than 90°, less than 180°
Straight angle	180°
Complementary angles	Two angles whose sum is 90°
Supplementary angles	Two angles whose sum is 180°

Circles

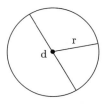

r = radius (plural *radii*)

d = diameter

c = circumference, the perimeter of a circle

$\pi \approx 3.14$

d = 2r

c = 2πr or πd

$A = \pi r^2$

Let's do a few examples.

EXAMPLE 1—

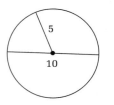

Find the area of a circle if the diameter is 10 feet.

If d = 10, then r = 5.

$A = \pi r^2 = \pi(5)^2 = 25\pi$

I and some teachers give this as an answer, while others want A = (3.14)25 = 78.5 square feet.

EXAMPLE 2—

Find the area and perimeter of a rectangle where the base is 5 meters and the height is 7 meters.

$p = 2b + 2h = 2(5) + 2(7) = 24$ meters

$A = bh = 5(7) = 35$ square meters

Let's try a more complicated one.

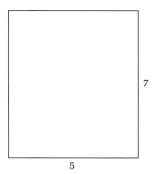

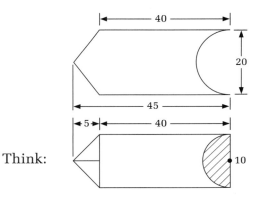

Think:

We have to think of it as:

a triangle   plus   a rectangle   minus   one half a circle

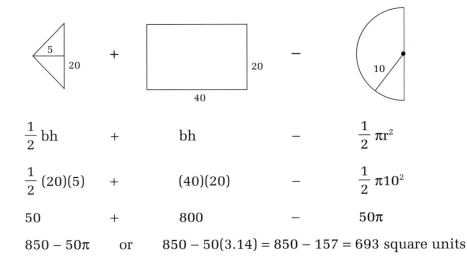

$\frac{1}{2} bh$ + bh − $\frac{1}{2} \pi r^2$

$\frac{1}{2} (20)(5)$ + $(40)(20)$ − $\frac{1}{2} \pi 10^2$

50   +   800   − 50π

$850 - 50\pi$ or $850 - 50(3.14) = 850 - 157 = 693$ square units

These sections are *very, very, very* important. Many students who are much more advanced forget some of these facts and get into trouble. Pleeease, learn them well.

The next section is also very important!!!

SOME DEFINITIONS, ADDITION, AND SUBTRACTION

As mentioned, there are relatively few definitions. But there are more that we need now.

Term—Any single collection of algebraic factors, which is separated from the next term by a plus or minus sign. Four examples of terms are $4x^3y^{27}$, x, $-5tu$, and 9.

A *polynomial* is one or more terms where all the exponents of the variables are natural numbers.

Monomials—Single-term polynomials: $4x^2y$, 3x, $-9t^6u^7v$.

Binomials—Two-term polynomials: $3x^2 + 4x$, $x - y$, $7z - 9$, $-3x + 2$.

Trinomials—Thrrreee-term polynomials: $-3x^2 + 4x - 5$, $x + y - z$.

Coefficient—Any collection of factors in a term is the coefficient of the remaining factors.

If we have 5xy, 5 is the coefficient of xy, x is the coefficient of 5y, y is the coefficient of 5x, 5x is the coefficient of y, 5y is the coefficient of x, and xy is the coefficient of 5. Whew!!!

Generally, when we say the word coefficient, we mean *numerical coefficient*. That is what we will use throughout the book unless we say something else.

Soooo, the coefficient of 5xy is 5. Also, the coefficient of −7x is −7. The sign is included.

The *degree* of a polynomial is the highest exponent of any one term.

EXAMPLE 1—

What is the degree of $-23x^7 + 4x^9 - 222$?

ANSWER

The degree is 9.

EXAMPLE 2—

What is the degree of $x^6 + y^7 + x^4y^5$?

ANSWER

The degree of the first term is 6; the second term is 7; the third term is 9 (= 4 + 5)

The degree of the polynomial is 9. The degree of x is 6 (the highest power of x). The degree of y is 7.

We will need the first example almost all of the time.

EXAMPLE 3—

Tell me about $5x^7 - 3x^2 + 5x$.

ANSWER

1. It is a polynomial because all the exponents are natural numbers.

2. It is a trinomial because it is 3 terms.

3. $5x^7$ has a coefficient of 5, a *base* of x, and an exponent (power) of 7.

4. $-3x^2$ has a coefficient of −3, base x, and exponent 2.

5. 5x has a coefficient of 5, a base of x, and an exponent 1.

6. Finally, the degree is 7, the highest exponent.

EXAMPLE 4—

Tell me about −x.

ANSWER

It is a monomial. The coefficient is −1. The base is x. The exponent is 1. The degree is 1.

−x really means $-1x^1$. The 1s are not written in. Write them in if you need to now. Later, you should not write them in.

In order to add or subtract, we must have like terms.

Like terms are terms with the exact letter combination *and* the same letters must have identical exponents.

x = x is called the reflexive law. An algebraic expression always equals itself.

We know x = x and abc = abc. Each pair are like terms.

a and a^2 are not like terms because the exponents are different.

x and xy are not like terms.

$2x^2y$ and $2xy^2$ are not like terms because $2x^2y = 2xxy$ and $2xy^2 = 2xyy$.

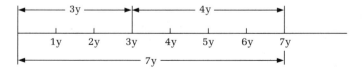

As pictured, 3y + 4y = 7y. Also, $7x^4 − 5x^4 = 2x^4$.

To add or subtract like terms, add or subtract their coefficients; leave the exponents unchanged.

Unlike terms cannot be combined.

EXAMPLE 5—

Simplify

$4x^2 + 5x + 6 + 7x^2 - x - 2$

ANSWER

$11x^2 + 4x + 4$

Letters in one variable are usually written highest exponent to lowest.

EXAMPLE 6—

Simplify

$4a + 9b - 2a - 6b$

ANSWER

$2a + 3b$

Terms are written alphabetically.

EXAMPLE 7—

Simplify

$5x + 7y - 5x + 2y$

ANSWER

$9y$. $5x - 5x = 0$ and is not written.

We can ignore the order of addition because of the associative law and the commutative law.

Commutative law $a + b = b + a$

$4x + 5x = 5x + 4x$

Associative law $a + (b + c) = (a + b) + c$

$(3 + 4) + 5 = 3 + (4 + 5)$

We will deal a lot more with minus signs in the next chapter.

After you are well into this book, you'll think these first pages were very easy. But some of you may be having trouble because the subject is so new. Don't worry. Read the problems over. Solve them yourself. Practice in your textbook. Everything will be fine!

PRODUCTS, QUOTIENTS, AND THE DISTRIBUTIVE LAW

Suppose we want to multiply x^5 times x^2

$$x^5(x^2) = (xxxxx)(xx) = xxxxxxx \qquad \text{or} \qquad x^5x^2 = x^7$$

Products

If the bases are the same, add the exponents:

$$x^m x^n = x^{m+n}$$

Different bases are not combined. Coefficients are multiplied.

EXAMPLE 1—

x^4x^5

ANSWER

x^9

EXAMPLE 2—

y^6y^8y

ANSWER

y^{15}

Remember, $y = y^1$.

EXAMPLE 3

$(3x^4y^5)(7xy^8)$

ANSWER

$21x^5y^{13}$

EXAMPLE 4

$(2^{18})(2^7)$

ANSWER

2^{25}

Base stays the same.

Order is alphabetical although order does not count because of the *commutative law* of multiplication and the *associative law* of multiplication.

Commutative law $ab = ba$

$3(7) = 7(3)$

$(4x^2)(5x^3) = (5x^3)(4x^2)$ $(=20x^5)$

Associative law $(ab)c = a(bc)$

$(2 \times 3)(5) = 2(3 \times 5)$ $(=30)$

$(4a \times 5b)(6d) = 4a(5b \times 6d)$ $(=120abd)$

EXAMPLE 5

Simplify

$4x^2(3x^3) + 5x(2x^2) - x(5x^4)$

SOLUTION

$12x^5 + 10x^3 - 5x^5 = 7x^5 + 10x^3$

Quotients

$$\frac{a}{b} = c \qquad \text{if } a = bc$$

$$\frac{12}{3} = 4 \qquad \text{because } 12 = 3(4)$$

THEOREM

Theorem: **A proven law.**

Division by 0 is not allowed.

Whenever I teach elementary algebra, this is one of the few theorems I prove because it is soooo important. Zero was a great discovery, in India in the 600s. Remember, Roman numerals have no zero. We must know why 6/0 has no meaning and 0/0 can't be defined.

PROOF

$$\frac{a}{0} \qquad \text{where } a \neq 0$$

Suppose a/0 = c. This means a = 0(c). But 0(c) = 0. So a = 0. Buuuut we assumed a ≠ 0. So a/0 is impossible. (7/0 is impossible).

$$\frac{0}{0}$$

Suppose 0/0 = c. This means 0 = 0(c). But c could be anything!!!!

This is called *indeterminate.* So 0/0 is not allowed. But 0/8 = 0 because 0 = 8(0).

When we are doing any divisions, we will assume that the denominators are not 0. $x^5/x^2 = x^3$ because $x^5 = x^2x^3$. Let us look at it a different way:

$$\frac{x^5}{x^2} = \frac{\overset{1}{\cancel{x}}\,\overset{1}{\cancel{x}}\, x\, x\, x}{\underset{1}{\cancel{x}}\,\underset{1}{\cancel{x}}} = x^3$$

Division

If the bases are the same, we subtract exponents:

$$\frac{a^m}{a^n} = a^{m-n}$$

EXAMPLE 1—

$$\frac{24x^5y^6z^7}{6x^2yz^6} = 4x^3y^5z \quad (x^{5-2}y^{6-1}z^{7-6}) \quad \text{and} \quad \frac{24}{6} = 4$$

EXAMPLE 2—

$$\frac{x^6}{x^6} = \frac{4}{4} = \frac{3 \text{ pigs}}{3 \text{ pigs}} = 1$$

EXAMPLE 3—

$$\frac{5x^3 + 7x^3}{2x} = \frac{12x^3}{2x} = 6x^2$$

EXAMPLE 4—

$$\frac{8x^6 + 20x^6}{2x^2} - \frac{8x^5}{2x}$$

$$\frac{28x^6}{2x^2} - \frac{8x^5}{2x}$$

Order of operations—
simplify top of fraction.

$$14x^4 - 4x^4$$

Order of operations—
divide, law of exponents.

$$10x^4$$

Combine like terms.

Distributive Law

We wind up this chapter with perhaps the most used law and perhaps the most liked law.

Distributive law $a(b + c) = ab + ac$

EXAMPLE 1—

$3(4a + 5b) = 12a + 15b$

EXAMPLE 2—

$5(4x + 5y + z) = 20x + 25y + 5z$

EXAMPLE 3—

$4x^3(7x^9 + 2x + 3) = 28x^{12} + 8x^4 + 12x^3$

EXAMPLE 4—

Multiply and simplify

$4(2a + 3b) + 6(5a + b) = 8a + 12b + 30a + 6b = 38a + 18b$

EXAMPLE 5—

$(4x + 7y) + (5x - 2y) = 1(4x + 7y) + 1(5x - 2y)$

$= 4x + 7y + 5x - 2y = 9x + 5y$

(You do not have to write the second step—just know it.)

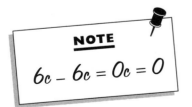

NOTE

$6c - 6c = 0c = 0$

EXAMPLE 6—

$(4a + 5b + 6c) + (a - 2b - 6c) = 4a + a + 5b - 2b + 6c - 6c$

$= 5a + 3b$

(Again, you do not have to write the second step.)

CHAPTER 2

INTEGERS PLUS MORE

Even though in a while you will look back at Chap. 1 as being very easy, for many of you Chap. 1 is not easy now. There is good news for you. Most of Chap. 2 is a duplicate of Chap. 1. The difference is that we will be dealing with the set of *integers.*

The *integers* are the set . . . −3, −2, −1, 0, 1, 2, 3, . . . or written 0, ±1, ±2, ±3, ...

±**7 means two numbers, +7 and −7.**

The *positive integers* is another name for the natural numbers.

x positive: x > 0, x is greater than zero.

The *nonnegative integers* is another name for the whole numbers.

x negative: x < 0, x is less than zero.

Even integers: 0, ±2, ±4, ±6, ±8 . . . .

Odd integers: ±1, ±3, ±5, ±7, . . . and so on.

3 means +3.

EXAMPLE 1

Graph the set −4, −1, 0, 3.

EXAMPLE 2—

Graph all even integers between −6 and 7. (*Between* means not including the end numbers.)

Now that we know what an integer is, we would like to add, subtract, multiply, and divide them.

ADDITION

For adding, you should think about money: + means gain and − means loss.

EXAMPLE 3—

7 + 5

Think (don't write)

(+7) + (+5)

Gain 7; gain 5 more.

ANSWER

12   or   +12

EXAMPLE 4—

−3 − 4

Think

(−3) + (−4)

Lose 3; lose 4 more.

ANSWER

−7

NOTE

You should think of this as an adding problem.

EXAMPLE 5

−7 + 3

Think

(−7) + (+3)

Lose seven; gain three: lose four.

ANSWER

−4

NOTE

−7 + 3 is the same problem as 3 − 7 (+3 + −7).

EXAMPLE 6

−7 + 9

Think

(−7) + (+9)

Lose seven; gain 9: up 2.

ANSWER

+2 or 2

NOTE

−7 + 9 is the same as 9 − 7 = (+9) + (−7) = 2.

You should read these examples until they make sense. Here are the rules in words:

Addition 1: If two (or all) of the signs are the same, add the numbers without the sign and put the sign that is in common.

Addition 2: If two signs are different, subtract the two numbers without the sign, and put the sign of the larger number without the sign.

EXAMPLE 7

$-7 - 3 - 2 - 4 - 1$

All the signs are negative.

ANSWER

-17

EXAMPLE 8

$-9 + 6$ or $6 - 9$

Signs are different. Subtract $9 - 6 = 3$. The larger number without the sign is 9. The sign of 9 is $-$.

ANSWER

-3

EXAMPLE 9

$8 - 4 - 9 + 3 + 5 - 7 - 10$

Add all the positives, $8 + 3 + 5 = 16$. Add all the negatives, $-4 - 9 - 7 - 10 = -30$. Thennnn $16 - 30 = -14$.

EXAMPLE 10

$4a - 5b - 7a - 7b$

Add like terms: $4a - 7a = -3a$; $-5b - 7b = -12b$.

ANSWER

$-3a - 12b$

Just like the last chapter!!!!

SUBTRACTION

Next is subtraction. We sort of avoided the definition of subtraction, but now we need it.

DEFINITION

Subtraction: $a - b = a + (-b)$

$6 - (+8) = 6 + (-8) = -2$

Important: There are only two real subtraction problems:

$-7 - (-3) = -7 \overset{+}{\ominus}(\overset{+}{\ominus}3) = -7 + (+3) = -4$

A number followed by a minus sign followed by a number in parentheses with a – sign in front of it.

$8 - (+2) = 8 \overset{+}{\ominus}(\overset{-}{\oplus}2) = +8 + (-2) = 6$

A number followed by a minus sign followed by a number in parentheses with a + sign in front.

All other problems should be looked at as adding problems, such as

$-4 - (5)$ is the same as $-4 - 5 = -9$

$-3 + (-7)$ is an adding problem; answer $= -10$

> **What we are doing is changing all subtraction problems to addition problems.**

> ***Only*** **one sign between is always adding.**

EXAMPLE 11—

$4a - (-5a) - (+3a) + (-8a) - 6a$

Subtract, change two signs; subtract, change two signs; add, no signs change; one sign, add, no signs change.

$4a + (+5a) + (-8a) + (-3a) + (-6a) = 9a + (-17a) = -8a$

The rest of the problems are just like addition.

MULTIPLICATION

The rules for multiplication, which are the same for division, will be shown by two simple patterns.

From 2 to 1 is down 1.

$$-1 \quad \begin{array}{c} (+2)(+3) = +6 \\ (+1)(+3) = +3 \end{array} \quad -3 \quad \text{Answer goes down 3.}$$

From 1 to 0 is down 1.

$$-1 \quad \begin{array}{c} \\ (0)(+3) = 0 \end{array} \quad -3 \quad \text{Answer goes down 3.}$$

From 0 to −1 is down 1.

$$-1 \quad \begin{array}{c} \\ (-1)(+3) = -3 \end{array} \quad -3 \quad \text{Product is −3, down 3.}$$

We just showed that a minus times a plus is a minus. By the *commutative* law, a plus times a minus is a minus. One more pattern:

$$-1 \quad \begin{array}{c} (+2)(-3) = -6 \\ (+1)(-3) = -3 \end{array} \quad +3 \quad \begin{array}{l} \text{Answer goes from −6} \\ \text{to −3. } \textit{Uppp } 3!!!! \end{array}$$

$$-1 \quad \begin{array}{c} \\ (0)(-3) = 0 \end{array} \quad +3 \quad \begin{array}{l} \text{Answer from −3 to 0,} \\ \text{again up 3.} \end{array}$$

$$-1 \quad \begin{array}{c} \\ (-1)(-3) = +3 \end{array} \quad +3 \quad \begin{array}{l} \text{From 0 to −1, down 1.} \\ \text{Answer up 3.} \end{array}$$

We just showed that a minus times a minus is a plus.

More generally, if the problem has only multiplications and divisions . . . *don't look at the plus signs at all:*

Odd *number of minus, answer is* **minus.**

Even *number of minus, answer is* **plus.**

EXAMPLE 12—

$$\frac{6(-2)(-3)(-4)}{(-12)(-4)(+1)}$$

5 minus signs; answer is minus.

ANSWER

−3

EXAMPLE 13—

$(-3a^4b^5c^6)(-10a^5b^{100}c^{1000})(5a^2b^3c)$

Determine sign first: 2 minus signs, +. Rest of the coefficient: $(3)(5)(10) = 150$; $a^{4 + 5 + 2} = a^{11}$; $b^{5 + 100 + 3} = b^{108}$; $c^{6 + 1000 + 1} = c^{1007}$.

Remember: **when you multiply, add the exponents, if the bases are the same.**

ANSWER

$+150a^{11}b^{108}c^{1007}$

Notice, big numbers do *not* make problems hard.

NOTE

$(a^m)^n = a^{mn}$

We'll do more of this later.

EXAMPLE 14—

$$\frac{(-3a^2)^4(-a^6)^3(4a)^3}{(2a^5)^5}$$

$$\frac{(-3a^2)(-3a^2)(-3a^2)(-3a^2)(-a^6)(-a^6)(-a^6)(4a)(4a)(4a)}{(2a^5)(2a^5)(2a^5)(2a^5)(2a^5)}$$

$(a^4)^3 = a^{12}$ because $(a^4)^3 = a^4a^4a^4 = a^{12}$.

Determine the sign: 7 minus signs—odd number. Answer is minus!!

Arithmetic trick: Always divide first, if possible. Cancelling is division!! It makes the work shorter, much shorter, or much much shorter.

$$\frac{3 \times 3 \times 3 \times 3 \times 4 \times 4 \times 4}{2 \times 2 \times 2 \times 2 \times 2} = 162$$

Remember: **when you divide, subtract the exponents, if the bases are the same.**

If you multiplied first, you would get $(3)(3)(3)(3)(4)(4)(4) = 5184$.

Then you would multiply out the bottom and get 32.

Lastly, you would have to long-divide 5184 by 32!!!! Cancelling makes this problem much easier.

Let's do the exponents: top is $a^{2 + 2 + 2 + 2 + 6 + 6 + 6 + 1 + 1 + 1} = a^{29}$; bottom is $a^{5 + 5 + 5 + 5 + 5} = a^{25}$; $a^{29}/a^{25} = a^{29 - 25} = a^4$. The answer is $-162a^4$. It takes longer to write it out than to do it.

EXAMPLE 14A—

We do inside the ( ) first.

$$\left(\frac{9x^7y^8}{3x^4y^2}\right)^2$$

Don't write this step if you don't have to.

$$= \left(\frac{9}{3}\frac{x^7}{x^4}\frac{y^8}{y^2}\right)^2$$

$$= (3x^3y^6)^2$$

$$= 9x^6y^{12}$$

EXAMPLE 14B—

Different exponents mean you must do outside exponents first.

$$\frac{(4x^3y^6)^2}{(2x^1y^3)^4}$$

16s cancel; y^{12}s cancel.

$$= \frac{16x^6y^{12}}{16x^4y^{12}}$$

$$= x^2$$

At this point, you should be doing pretty well. If not, please read the intro again. Work at this. If you do, you will be terrific!!!!

RATIONAL NUMBERS

Informally, we call these numbers fractions, but that is technically not correct. There are two definitions of rational numbers:

Definition 1: Any number a/b where a and b are integers; b cannot be 0.

Definition 2: Any number that can be written as a repeating decimal.

$-1/6 = -0.166666 \ldots = -0.1\overline{6}$ is a rational number.
$1/4 = 0.25000000$ is a rational number, buuut $\pi/6$ is not a rational number even though it is a "fraction."

We now repeat what we've done for natural numbers and integers. Let's give a few examples.

EXAMPLE 15—

If $x = 1/2$ and $y = -1/4$, evaluate $xy^2 - 6x - 2$.

$$xy^2 - 6x - 2 = \left(\frac{1}{2}\right)\left(-\frac{1}{4}\right)^2 - 6\left(\frac{1}{2}\right) - 2$$

$$= \frac{1}{2}\left(\frac{1}{16}\right) - 6\left(\frac{1}{2}\right) - 2$$

$$= \frac{1}{32} - 3 - 2 = -5 + \frac{1}{32} \quad \text{or} \quad -\frac{159}{32}$$

EXAMPLE 16—

Simplify

$$\frac{3}{8}x - \frac{1}{6}y - \frac{1}{16}x + x - \frac{1}{3}y$$

ANSWER

$$\frac{3}{8}x - \frac{1}{16}x + 1x - \frac{1}{6}y - \frac{1}{3}y = \frac{21}{16}x - \frac{1}{2}y$$

If this is the first section you are having trouble with, then your trouble is with fractions. So practice!!! In fact, let's talk about practice. You should practice until you know a topic well. Sometimes that means doing 2 problems; sometimes 20 problems; and, unfortunately, sometimes 100 problems. Each topic is different. When you know the topic, you've practiced enough.

DIVISION

Interestingly, in arithmetic most students have the most problems with division, but in algebra most students like division best. The next example is a second look at exponents.

EXAMPLE 17—

We know that $x^7/x^2 = x^5$. What about x^3/x^8?

$$\frac{x^3}{x^8} = \frac{\overset{1\ 1\ 1}{\cancel{x}\,\cancel{x}\,\cancel{x}}}{\underset{1\ 1\ 1}{\cancel{x}\,\cancel{x}\,\cancel{x}\,x\,x\,x\,x\,x}} = \frac{1}{x^5}$$

The complete rule for division is as follows:

The base remains the same.

$$\frac{x^m}{x^n} = x^{m-n} \qquad \text{if } m > n$$

$$\frac{x^m}{x^n} = \frac{1}{x^{n-m}} \qquad \text{if } n > m$$

In simple English, in division, subtract the larger exponent minus the smaller; the term goes wherever the larger exponent was.

EXAMPLE 18—

$$\frac{18a^4b^5c^6d^7e^8}{9a^7b^2c^6d^8e} = \frac{2b^{5-2}e^{8-1}}{a^{7-4}d^{8-7}} = \frac{2b^3e^7}{a^3d}$$

$$\frac{18}{9} = 2 \qquad \frac{c^6}{c^6} = 1$$

EXAMPLE 19—

$\frac{4}{10}$ **is reduced.**

a. $\dfrac{4x^6y^3}{10x^2y^9} = \dfrac{2x^4}{5y^6}$

When the whole top cancels, you must put the 1 (one) on top.

b. $\dfrac{2x^3}{6x^5} = \dfrac{1}{3x^2}$

Base stays the same.

c. $\dfrac{7^5}{7^{11}} = \dfrac{1}{7^6}$

SHORT DIVISION

This is the opposite of adding fractions. We know

$$\frac{a}{c} + \frac{b}{c} = \frac{a+b}{c}$$

$$\frac{5}{7} + \frac{3}{7} = \frac{8}{7}$$

The opposite is

$$\frac{a+b}{c} = \frac{a}{c} + \frac{b}{c}$$

EXAMPLE 20—

$$\frac{8x^6 + 12x^5}{2x^3} = \frac{8x^6}{2x^3} + \frac{12x^5}{2x^3} = 4x^3 + 6x^2$$

Let's try a longer one. Remember, longer is not harder. I will tell you when things get harder.

EXAMPLE 21—

$$\frac{12x^8 - 14x^7 + 4x^4 + 2x}{4x^4}$$

$$\frac{12x^8}{4x^4} - \frac{14x^7}{4x^4} + \frac{4x^4}{4x^4} + \frac{2x}{4x^4} = 3x^4 - \frac{7x^3}{2} + 1 + \frac{1}{2x^3}$$

NOTES

1. The 1 must be put in since it is added.

2. If there are 4 unlike terms on the top of the fraction at the start, there must be 4 terms at the end.

3. There is another way to do this problem, as we will learn later.

4. Most students like this kind of problem, but *many* forget this problem. Please don't forget!!!!

DISTRIBUTIVE LAW

Finally, we will do the *distributive law*. It probably doesn't belong here, but it is needed, and again, it is not too hard.

Distributive law $a(b + c) = ab + ac$

If a term is on the outside of the parentheses and inside the parenthetical terms are added or subtracted, every term on the inside is multiplied by the outside term.

EXAMPLE 22—

a. $4(3x - 5y - 6z) = 12x - 20y - 24z$

b. $5x(6x - 7y - 8z) = 30x^2 - 35xy - 40xz$

c. $6a^2(7a^3 - 5a^2 - a + 4) = 42a^5 - 30a^4 - 6a^3 + 24a^2$

d. $-4x^5y^7(-2x^{77}y^{32} + 9xy^3) = +8x^{82}y^{39} - 36x^6y^{10}$

EXAMPLE 23—

Multiply and simplify

a. $4(3x - 5) + 6(2x + 5)$

b. $5(2x - 5) - 7(2x - 8)$

c. $2x(5x - 7) + 3x(4x^2 - 2)$

d. $(2a + b - c) + (5c - b) - (4b - 2a)$

ANSWERS

a. $12x - 20 + 12x + 30 = 24x + 10$

b. $10x - 25 - 14x + 56 = -4x + 31$

c. $10x^2 - 14x + 12x^3 - 6x = 12x^3 + 10x^2 - 20x$

d. $1(2a + b - c) + 1(5c - b) - 1(4b - 2a)$

$\quad = 2a + b - c + 5c - b - 4b + 2a$

$\quad = 4a - 4b + 4c$

Careful: $-7(-8) = +56$.

Combine like terms.

This step not necessary once you know it. In fact, try to do as many steps in your head as possible, once you improve.

Now let's get to the chapter almost everyone likes!!!!

CHAPTER 3

FIRST-DEGREE EQUATIONS

We want to solve first-degree equations. Most of you will really like this topic once you are finished with this chapter. Imagine that an equation is like a balance scale.

We can rearrange objects on either side (mathematically, we combine like terms, use the distributive law . . .) and the scale stays balanced. But . . . if we add something to the left, we must add exactly the same thing to the right; similarly, if we subtract from one side, multiply one side, divide one side, we must do exactly the same to the other side.

An *equation* is an algebraic expression with a left side, a right side, and an equal sign in between.

EXAMPLE 1—

$4xy - 7y^3 = 5xz + m^{88} - 12$

left side = right side

OK, OK, we are not going to deal with this kind of equation. Let's look at some real ones.

A *solution* (root) of an equation is a number or expression that balances the sides.

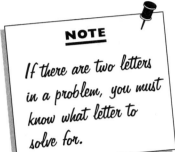

EXAMPLE 2—

Given $5x - 2 = 2x - 23$, 5 is *not* a root, while -7 *is* a root.

$5(5) - 2 \overset{?}{=} 2(5) - 23$ $5(-7) - 2 \overset{?}{=} 2(-7) - 23$

$\qquad 23 \ne -13$ $\qquad\qquad -37 = -37$

5 is *not* a root -7 *is* a root

EXAMPLE 3—

$6x - 2a = 4x - 10a$ has a root $x = -4a$.

$6(-4a) - 2a \overset{?}{=} 4(-4a) - 10a$

$-26a = -26a$

Soooo, $x = -4a$ is a root.

There are 3 basic types of equations we will deal with.

1. *Contradiction*—An equation with no solution or a false equation.

EXAMPLE 4—

$9 + 7 = 4$

$x = x + 1$

You cannot add one to a number and get the same number.

We will do only a few of these.

2. *Identity*—An equation that is true for all values it is defined for.

EXAMPLE 5—

$2x + 3x = 5x$

No matter what value we substitute, 2 times a number plus 3 times that number will always be 5 times that number.

EXAMPLE 6—

$$\frac{5}{x} + \frac{6}{x} = \frac{11}{x}$$

It is not defined for $x = 0$ (remember, you can't divide by 0), but for any other number $5/x + 6/x = 11/x$.

We will not deal with this kind of equation very much right now, but we will later (much later). About 95 percent, maybe more, of our equations will be conditional equations.

3. *Conditional*—An equation that is true for some value but not all values.

Examples 2 and 3 are examples of first-degree (*linear*) conditional equations. A linear equation that is not a contradiction or an identity has one solution.

EXAMPLE 7—

$x^2 - 2x = 8$ has two roots, 4 and −2.

$$(4)^2 - 2(4) = 8 \qquad \text{and} \qquad (-2)^2 - 2(-2) = 8$$

Such equations are called *Quadratic* (second-degree) equations. If they are not identities or contradictions, they will always have two roots. We shall study them in great detail several times later.

SOLVING LINEAR EQUATIONS

It is gently snowing outside with a blizzard coming, a wonderful day to write a section that most of you will like very much.

EXAMPLE 1—

Solve for x: $4x + 8 = 10x - 16$.

We would like to know the steps to use so that we can solve this equation.

Opposite (additive inverse):
$a + (-a) = (-a) + a = 0.$

The opposite of 3 is −3 because $3 + (-3) = 0$.

The opposite of −6b is 6b because $-6b + 6b = 6b + (-6b) = 0$.

The opposite of 0 is 0.

1. Fractions cause the most problems. So we get rid of them at once.

2. The purpose is to make all equations look the same, so that you get very good very fast.

1. Multiply each term by the least common denominator (LCD).

2. If the x terms are only on the right, switch the sides.

3. Multiply out all parentheses, brackets, and braces.

4. On each side combine like terms.

5. Add the opposite of the x terms on the right to each side.

6. Add the opposite of the non-x terms on the left to each side.

If you notice, 7 is missing. We will get a 7 later.

8. Divide each side by the whole coefficient of x, including the sign.

We will go down each step for each example. Let us rewrite Example 1.

EXAMPLE 1—

Solve for x: $4x + 8 = 10x - 16$.

1. Multiply each term by the LCD. We can't do this because there are no fractions.

$4x + 8 = 10x - 16$

2. If the x terms are only on the right, switch the sides. We can't do this because there are x terms on both sides.

3. Multiply out all parentheses, brackets, and braces. None here.

$$4x + 8 = 10x - 16$$

4. On each side, combine like terms. On the left, 4x + 8 has no like terms; on the right, 10x − 16 has no like terms. Pretty easy so far . . . but there is a step 5.

$$-10x \quad = -10x$$
$$-6x + 8 = \quad -16$$

5. Add the opposite of the x terms on the right to both sides. The opposite of 10x is −10x.

$$-8 = \quad -8$$
$$-6x \quad = \quad -24$$

6. Add the opposite of the non-x terms on the left to each side. The opposite of +8 is −8.

$$\frac{-6x}{-6} = \frac{-24}{-6}$$

8. Divide each side by the whole coefficient of x, including the sign.

$$x \quad = \quad 4$$

The root is x = 4.

To check, substitute x = 4 in the original equation—
always, always the original in case of error.

$$4(4) + 8 \overset{?}{=} 10(4) - 16$$

$$24 = 24$$

Yes, it is correct!!!!

It is not necessary to memorize these steps. You should just keep them in front of you. With practice you will know them. Not only that, but you will be able to skip some steps and change the order. You will know without anyone telling you. Also do each problem in the book before you do new ones. Be sure to understand each step. It will really help you. Let's try some more.

EXAMPLE 2

Solve for y: $12 - 2(3y - 4) = 4[7 - y]$.

$$12 - 2(3y - 4) = 4[7 - y]$$

There are no fractions and the y terms are on both sides, but there is a step 3.

3. Multiply out all ( ) and [ ].

$$12 - 6y + 8 = 28 - 4y$$

4. We can combine like terms on the left. (We always write the y terms first, since we want to make all the problems look the same.)

$$-6y + 20 = \qquad -4y + 28$$

5. Add the opposite of the y term on the right to both sides.

$$+4y \qquad = \qquad +4y$$
$$-2y + 20 = \qquad +28$$

6. Add the opposite of the non-y terms on the left to each side. When we write −20 = −20 or, in general, b = b, we are using the *reflexive law,* which says that anything equals itself.

$$-20 = \qquad -20$$
$$-2y \qquad = \qquad 8$$

8. Divide each side by the whole coefficient of y, including the sign. In some books, they say multiply by its *reciprocal* or *multiplicative inverse*—if a ≠ 0, a × (1/a) = (1/a) × a = 1. Reciprocal of −2 is 1/−2. Reciprocal of 3/4 is 4/3.

$$\frac{-2y}{-2} = \frac{8}{-2}$$
$$y = -4$$

Even if you are pretty good at this right away and can do steps in your head, practice writing all the steps down, *carefully,* lining up the equal signs.

To check:

$$12 - 2(3y - 4) = 4[7 - y] \qquad \text{for } y = -4$$

$$12 - 2(3(-4) - 4) \overset{?}{=} 4[7 - (-4)]$$

$$12 - 2(-16) \overset{?}{=} 4[11]$$

$$12 + 32 \overset{?}{=} 44$$

$$44 = 44$$

EXAMPLE 3

Solve for x:

$$\frac{2}{3} = \frac{x}{4} + \frac{7x - 5}{6}$$

$$\frac{2}{3} = \frac{x}{4} + \frac{(7x - 5)}{6}$$

$$\frac{12}{1} \times \frac{2}{3} = \frac{12}{1} \times \frac{x}{4} + \frac{12}{1} \times \frac{(7x - 5)}{6}$$

$$8 = 3x + 2(7x - 5)$$

$$3x + 2(7x - 5) = 8$$

$$3x + 14x - 10 = 8$$

$$17x - 10 = 8$$

$$+10 = +10$$

$$\frac{17x}{17} = \frac{18}{17}$$

$$x = \frac{18}{17}$$

1. Multiply each term by the **LCD, 12. Play it safe!!** If any fraction has more than one term on top (or later on the bottom), put ( ) around it. Multiply each term by **LCD/1** (12/1) so you know the 12 is on top.

Notice how much nicer the problem looks without fractions!!!

2. If the x terms appear only on the right, switch the sides. This step has a name, the *symmetric law.* It says that if a = b, then b = a (we can always switch sides).

3. Multiply out all ( ).

4. Combine like terms on the left.

6. Add +10 to both sides

8. Divide each side by 17.

Sometimes the check is worse than doing the problem. This is definitely the case here.

$$\frac{2}{3} \stackrel{?}{=} \frac{\left(\dfrac{18}{17}\right)}{4} + \frac{7\left(\dfrac{18}{17}\right) - 5}{6}$$

$$\stackrel{?}{=} \frac{\dfrac{18}{17} \times \dfrac{17}{1}}{\dfrac{4}{1} \times \dfrac{17}{1}} + \frac{\dfrac{7}{1}\left(\dfrac{18}{17}\right) - \dfrac{5}{1}\left(\dfrac{17}{17}\right)}{\dfrac{6}{1}}$$

$$\stackrel{?}{=} \frac{18}{68} + \frac{\dfrac{126}{17} - \dfrac{85}{17}}{\dfrac{6}{1}}$$

$$\stackrel{?}{=} \frac{9}{34} + \frac{\dfrac{41}{17}\dfrac{17}{1}}{\dfrac{6}{1}\dfrac{17}{1}}$$

$$\stackrel{?}{=} \frac{9}{34}\frac{3}{3} + \frac{41}{102}$$

$$\stackrel{?}{=} \frac{27}{102} + \frac{41}{102}$$

$$\stackrel{?}{=} \frac{68}{102}$$

$$\frac{2}{3} = \frac{2}{3}$$

Whew! ALGEBRA is more fun than arithmetic, at least for me!!!

Solve for z:

EXAMPLE 4

$$\frac{2z - 3}{5} = \frac{4z - 7}{9}$$

$$\frac{(2z - 3)}{5} = \frac{(4z - 7)}{9}$$

$$5(4z - 7) = 9(2z - 3)$$

$$20z - 35 = \quad 18z - 27$$

$$-18z \qquad = -18z$$

$$2z - 35 = -27$$

$$+35 = +35$$

$$\frac{2z}{2} = \frac{8}{2}$$

$$z = 4$$

1. In the case of 2 fractions, in order to clear fractions we can *cross multiply*. Because . . . a/b = c/d if ad = bc or bc = ad. 4/6 = 6/9 because 4(9) = 6(6). We could have written 9(2z − 3) = 5(4z − 7).

No step 2. Step 3, multiply out ( ).

No step 4. But the last 3 steps are in most of our problems, steps 5, 6, and 8: add the opposite of the z term on the right to each side; add the opposite on the non-z term on the left to each side; and divide each side by the whole coefficient of z.

To check:

$$\frac{2(4) - 3}{5} \stackrel{?}{=} \frac{4(4) - 7}{9}$$

Because 5/5 = 9/9, the problem checks. Some checks are nice.

EXAMPLE 5—

Solve for t: $3t(t - 4) = 2(t - 7) + 3t^2$.

This doesn't look like a linear equation, but it will shortly reduce to a linear equation.

3. Multiply out all ( )

$$3t(t - 4) \quad = 2(t - 7) \ + 3t^2$$

5. Add the opposite of the t terms on the right $(-2t - 3t^2)$ to each side.

$$3t^2 - 12t = \ 2t - 14 + 3t^2$$
$$-3t^2 - \ 2t = -2t \qquad - 3t^2$$

8. Divide each side by -14. If the t^2 term did not drop out, we could not finish the problem now. Later we will. It is called a _quadratic_ equation.

$$\frac{-14t}{-14} = \frac{-14}{-14}$$

$$t = 1$$

To check:

$$3(1)((1) - 4) \stackrel{?}{=} 2((1) - 7) + 3(1)^2$$

Because $-9 = -9$, the problem checks.

EXAMPLE 6—

Solve for x:

$$4x + 6x + 3 = 10x + 27$$

4. Combine like terms

$$10x + 3 = 10x + 27$$

5. Add $-10x$ to each side

$$-10x = -10x$$

$$3 = 27$$

But 3 never equals 27. There is no solution to this problem. It is a _contradiction_ and does not have a solution. If we got 5 = 5, this would always be true, an _identity_.

EXAMPLE 7—

$9ax + 2ab - 12a^3 = 12ax + 20ab$

Now don't panic!!! We will use the same steps. And there usually is a lot less arithmetic. I like these.

We must always be told what letter to solve for. We will solve for x.

$$9ax + 2ab - 12a^3 = 12ax + 20ab$$

$$-12ax \qquad\qquad = -12ax$$

$$-3ax + 2ab - 12a^3 = 20ab$$

$$-2ab + 12a^3 = -2ab + 12a^3$$

$$\frac{-3ax}{-3a} = \frac{18ab}{-3a} + \frac{12a^3}{-3a}$$

$$x = -6b - 4a^2$$

No steps 1–4. Add the opposite of the x terms on the right to each side (step 5).

6. Add the opposite of the non-x terms on the left to each side (line up like terms).

8. Divide each term by the whole coefficient of x, which is −3a.

Not too bad.

Let's check. Boy, is it snowing now!!!! $x = -6b - 4a^2$ in the equation $9ax + 2ab - 12a^3 = 12ax + 20ab$ gives us

$$9a(-6b - 4a^2) + 2ab - 12a^3 \stackrel{?}{=} 12a(-6b - 4a^2) + 20ab$$

$$-54ab - 36a^3 + 2ab - 12a^3 \stackrel{?}{=} -72ab - 48a^3 + 20ab$$

$$-52ab - 48a^3 = -52ab - 48a^3$$

Problem checks. The check isn't too bad, either.

Sometimes, you have a formula that you have to solve for a different letter.

The area of a trapezoid $A = \frac{1}{2}h(b_1 + b_2)$, one-half the height times the sum of the bases.

EXAMPLE 8—

Solve for b_1:

1. **Multiply both sides by 2.**

$$A = \tfrac{1}{2}h(b_1 + b_2)$$

2. **Switch sides.**

$$2A = h(b_1 + b_2)$$

3. **Distributive law.**

$$h(b_1 + b_2) = 2A$$

6. **Get hb_2 to the right.**

$$hb_1 + hb_2 = 2A$$

$$-hb_2 = \qquad -hb_2$$

8. **Divide both sides by h.**

$$hb_1 = 2A - hb_2$$

$$b_1 = \frac{2A - hb_2}{h} \text{ or } \frac{2A}{h} - b_2$$

CHAPTER 4

PROBLEMS WITH WORDS

WHY SO MANY STUDENTS HAVE PROBLEMS ON THE SAT

One of the reasons so many of you have trouble in math, especially on the SAT, is the lack of problems with words. These problems help your brain translate algebra into English. In your future, when people give you problems—and they will—this skill will allow you to understand the problems better and help you solve them.

In your real life, problems will often be much more difficult.

Let's begin with some basic problems.

BASIC PROBLEMS

EXAMPLE 1

Two more than four times a number is twenty-eight less than ten times the number. Find the number.

Let's take a look at the pieces: "two more than four times a number."

We let x equal the number. Going back to Chap. 1, two more than four times a number: 4x + 2.

Plus in any order.

Less than reverses.

The second part: "twenty-eight less than ten times the number": $10x - 28$.

The last part: "is." The verb to be, is, am, was, were, will be, most times is the equal sign.

Let us go over: two more than four times a number is twenty-eight less than ten times the number. Now we solve:

$$
\begin{aligned}
4x + 2 &= 10x - 28 \\
-10x &= 10x \\
-6x + 2 &= -28 \\
-2 &= -2 \\
\frac{-6x}{-6} &= \frac{-30}{-6} \\
x &= 5
\end{aligned}
$$

The check is easy: $4(5) + 2 = 10(5) - 28$ because $22 = 22$. The reading takes time. Sometimes you have to read and reread the problems.

EXAMPLE 2—

In Springfield, the election for town mayor was close. The winner received 20 more votes than the loser. If 4800 votes were cast, how many votes did each get?

In a problem like this, you try to let x equal the smaller number (or the smallest if there are more than two). The loser received x votes. The winner received 20 more, or $x + 20$. If you add the winner and loser, you get 4800 votes: $x + x + 20 = 4800$.

Solving, we get $x = 2390$. But this isn't the only answer. 2390 is the loser's total. The winner gets $x + 20 = 2390 +$

20 = 2410. You must read the problem. Sometimes the problem asks for winner or loser or both.

EXAMPLE 3—

The difference between two numbers is one. The sum is 12. Find the numbers.

There are two ways (at least) to do this problem.

x = smaller number   x + 1 = the larger   their sum = 12
x + x + 1 = 12

Solving, we get x = 5½. So x + 1 = 6½. Orrrr

x = larger number   x − 1 = smaller   x + x − 1 = 12

Solving, x = 6½ and x − 1 = 5½.

There may be many ways to do the problem. Be creative. Notice that *number* does *not* necessarily mean *integer!!!!*

EXAMPLE 4—

A 45-inch board is cut into a ratio of 2 to 3. Find the length of each piece.

If the ratio is 2 to 3, we can let one piece = 2x and the other = 3x.

2x + 3x = 45. Solving, we get x = 9.

One piece is 2x = 2(9) = 18. The other piece is 3x = 3(9) = 27.

NOTES

1. Ratios have two common notations: 2:3 (read: "two to three") or 2/3.

2. We will do two more problems now, and a little more later, on ratios.

EXAMPLE 5—

In a triangle, the ratio of the angles is 3:4:5. Find the angles.

As in Example 4, the angles will be 3x, 4x, and 5x.

3x + 4x + 5x = 180, since the angles of a triangle add up to 180°.

Solving, x = 15, and the angles are 45°, 60°, and 75°.

EXAMPLE 6—

A fraction reduces to ⅔. If 5 is subtracted from the numerator and 7 is added to the denominator, the new fraction equals ½.

Find the original numerator and denominator.

We can let 2x = old numerator; the new one is 2x − 5.

The old denominator = 3x; the new one is 3x + 7.

The equation is:

$$\frac{2x - 5}{3x + 7} = \frac{1}{2}$$

Solving, we cross multiply and get 2(2x − 5) = 1(3x + 7). x = 17. The original fraction 2x/3x was 34/51.

Oh, let's check. 34 − 5 = 29 (new numerator). 51 + 7 = 58 (new denominator). 29/58 = ½, as promised.

NOTES

1. The checks of almost all word problems are not too hard.

2. You do not have to do hundreds of these problems to get good, but you must do more than a few, more than zero.

Let's do some other types.

Consecutive Integer Problems

If students had a favorite word problem, this would be it.

EXAMPLE 1

The sum of three consecutive integers is 93. Find them.

Recall that the integers are . . . −3, −2, −1, 0, 1, 2, 3, ... If we are talking about consecutive integers, they differ by 1.

Let x equal the first consecutive integer; x + 1 equal the second consecutive integer; and x + 2 equals the third consecutive integer. *Sum* means *add*. The equation is:

x + x + 1 + x + 2 = 93

3x + 3 = 93

3x = 90

Soooo, x = 30. That isn't the only answer. We are asked to find *all*. x = 1 = 31 and x + 2 = 32. The answers are 30, 31, and 32. Notice that the check, the sum is 93, is very, very easy.

EXAMPLE 2

Find four consecutive even integers so that the sum of the smallest two is two less than three times the largest.

Even integers are . . . −8, −6, −4, −2, 0, 2, 4, 6, ... The difference between consecutive even integers is 2. Let x equal the first consecutive even integer (the smallest); x + 2, the next; x + 4, the next; and x + 6 the largest. The sum of the two smallest is x + x + 2; 2 less than 3 times the largest is 3(x + 6) − 2.

The equation is:

$x + x + 2 = 3(x + 6) - 2$

$2x + 2 = 3x + 16$

$x = -14$

Remember, negatives are okay.

$x + 2 = -12$

$x + 4 = -10$

$x + 6 = -8$

The answers are -14, -12, -10, and -8. Not too bad, are they??

EXAMPLE 3

Find three consecutive odd integers such that the square of the smallest added to one more than the largest is the same as the product of the smallest and middle numbers.

Odd integers are . . . $-5, -3, -1, 1, 3, 5, 7, \ldots$. The difference between consecutive odds is also 2. We do the even and odd integer problems exxxxactly the same way. $x =$ odd integer (the smallest); $x + 2 =$ next consecutive odd integer (the middle); $x + 4 =$ largest. $x^2 =$ square of the smallest; one more than the largest $= x + 4 + 1$; product (answer in multiplication) of the smallest and the middle $= x(x + 2)$.

$x^2 + x + 4 + 1 = x(x + 2)$

$x^2 + x + 5 = x^2 + 2x$

x^2's cancel $x + 5 = 2x$

$x = 5$

The answers are 5, 7, and 9. As I said, most students like this kind of problem.

Geometry Problems

These problems require remembering certain facts from geometry. In addition, it may help to draw pictures. In fact, whenever you can draw pictures to help you, you absolutely, positively, should. A 6-inch ruler should be used to make the picture look reasonably nice.

EXAMPLE 1—

The vertex angle of an isosceles triangle is 6° more than the base angle.

Find the angles.

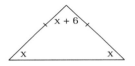

We know that an isosceles triangle has two equal base angles. Let x = base angle; x + 6 = vertex angle. We recall that the sum of the angles of a triangle is 180°. Soooo, according to the picture:

x + x + x + 6 = 180°

3x = 174°

x = 58°

x + 6 = 64°

The angles are 58°, 58°, and 64°.

EXAMPLE 2—

The supplement of an angle is three times its complement. Find the angle.

Okay, okay, not every geometry problem has a picture. Let x equal the angle; 90 − x is its complement and 180 − x is its supplement. Recall that *complements* are two angles whose sum is 90°, and *supplements* are two angles whose sum is 180°. Okay, let's explain a little more:

If the angle is 30°, the supplement is 180° − 30° = 150°.

If the angle is 71°, the supplement is 180° − 71° = 109°.

If the angle is x, the supplement is 180° − x.

Similarly, the complement of x is 90° − x.

The equation is:

180 − x = 3(90 − x)

180 − x = 270 − 3x

2x = 90

x = 45°

EXAMPLE 3

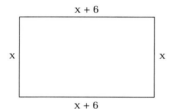

The length of a rectangle is 6 feet more than its width. The perimeter is 52 feet. Find the dimensions.

Let x = width (smaller one); x + 6 = length. *Perimeter* means add up all the sides. According to the picture:

x + x + x + 6 + x + 6 = 52

4x + 12 = 52

4x = 40

x = 10

x + 6 = 16

The problems in the next group appear to be very different, but they are all done basically the same way. We will use a chart because it is the easiest way to look at the problems. However, you should understand the chart. Understanding is allllwayssss important. Not all the boxes will be filled in on all the problems.

Mixtures, Coins, Percentages, and Interest

EXAMPLE 1—

How many pounds of peanuts selling at 60¢ a pound must be mixed with 12 pounds of walnuts selling at 90¢ a pound to give a mixture at 70¢ a pound? Can you imagine anybody actually doing this? Oh well, let's solve it.

The *theory* is that the cost per pound times the number of pounds is the total cost (80¢ per pound × 7 pounds = cost $5.60).

	Cost/pound	×	pounds	=	total cost
Peanuts	60		x		60x
Walnuts	90		12		90(12)
Mixture	70		x + 12		70(x + 12)

We let x = pounds of peanuts, the unknown in the problem. Since there are 12 pounds of walnuts, a lot of walnuts, the mixture must have x + 12 pounds. We multiply the first two columns. The equation is the cost of the peanuts plus the cost of the walnuts is the total cost of everything.

$60x + 1080 = 70(x + 12)$

$60x + 1080 = 70x + 840$

$10x = 240$

$x = 24$ pounds

NOTE

In many problems, the first two in a column add up to the third (mixture). The last column gives the equation. The first two (or sometimes more than two) add up to the bottom. Notice also that the decimal point isn't needed in this type of problem. Everything is done in pennies.

EXAMPLE 2—

We have nickels and dimes, 30 coins in all, totaling $2.60. How many of each coin?

THEORY

Value of a coin times number of coins equals total amount of money. 7 quarters are $7 \times 25¢ = \$1.75$.

	Value per coin	$\times$	number of coins	$=$	total value
Nickels	5		x		5x
Dimes	10		30 − x		10(30 − x)
Mixture	—		30		260

Let x equal the number of nickels. Since the total is 30, the other component (dimes) equals the total minus the number of nickels, or 30 − x. Again we multiply across and we get the equation, the value of nickels plus the value of dimes equals the total value in pennies.

$5x + 10(30 - x) = 260$

$5x + 300 - 10x = 260$

$-5x = -40$

$x = 8$ nickels

$30 - x = 22$ dimes

EXAMPLE 3—

How many gallons of 30% alcohol must be mixed with 80% alcohol to give 16 gallons of 60% alcohol?

THEORY

Number of gallons of substance × % alcohol = amount of alcohol.

12 gallons of 35% alcohol = 12 × 0.35 = 4.20 gallons of alcohol . . . except . . . it is easier to do the problem without changing to a decimal.

	Gallons substance	×	% alcohol	=	amount of alcohol
Substance A	x		30		30x
Substance B	16 − x		80		80(16 − x)
Mixture	16		60		16(60)

$30x + 80(16 - x) = 960$

$30x + 1280 - 80x = 960$

$-50x = -320$

$x = 6.4$

6.4 gallons 30% alcohol and 16 − 6.4 = 9.6 gallons 80% alcohol.

NOTES

1. The answer does not have to be an integer.

2. Notice how similar this problem is to the coin problem in Example 2.

3. Pure alcohol is 100% alcohol.

4. Water is 0% alcohol.

EXAMPLE 4—

An amount of money is invested at 6% and $800 more is invested at 5%. If we have $260 in interest, how much was invested at each rate?

NOTE

As some of you may wonder, this is simple interest, where the time is one year. Later,* much later, we will deal with time not one year and compound interest. It is not too difficult, but math does not go from easy to hard, but from topic to topic. Compound interest does not belong here.

THEORY

(Not much theory in any of the cases): Principal (amount invested) times rate equals interest. $20,000 at 7% = $20,000 \times 0.07 = $1400.

	Principal	$\times$	rate	$=$	interest
Amount A	x		.06		.06x
Amount B	x + 800		.05		.05(x + 800)
Mixture	—		—		260

Let x = smaller amount at 6%; x + 800 = amount at 5%.

$0.06x + 0.05(x + 800) = 260$

$6x + 5(x + 800) = 26{,}000$ Multiply by 100.

$6x + 5x + 4000 = 26{,}000$

$11x = 22{,}000$

$x = 2000$

$x + 800 = 2800$

ANSWER
$2000 at 6% and $2800 at 5%.

AGE PROBLEMS

Another popular type of problem is age problems.

EXAMPLE 1—

Suppose you are x years old.

a. In three years you will be x + 3 years old.

b. Four years ago, you were x − 4 years old.

c. Five times your age is 5x.

d. Half your age is (½)x.

*See *Precalc with Trig for the Clueless*.

EXAMPLE 2

a years ago, you were b years old; in c years, how old will you be?

The secret of doing these kinds of problems is to find your age now. Now you are a + b years old. In c years, you will be a + b + c years old.

EXAMPLE 3

Randy is four times Sandy's age. In five years, Randy will be three times Sandy's age. How old are they now?

Most of the time, we let x = the smaller quantity. Sandy is x years old now. Randy is 4x years old now. In five years, Sandy will be x + 5 years and Randy will be 4x + 5 years old. The equation then rights itself.

In five years, Randy will be three times Sandy's age.

$$4x + 5 = 3(x + 5)$$

Solving, Sandy's age now, x, is 10 years. Randy's age, 4x, is 40 years old.

To check, in five years Randy will be 45, three times Sandy's age, 15.

NOTE

This is a past and future (?) SAT type question.

Distance Problems

The last kind of problem, the distance problem, is the only truly important problem in terms of mathematics because distance is a physical quantity.

If we drive 30 miles per hour for 3 hours, we go 90 miles. Or in words: rate times time equals distance; or in letters, $r \times t = d$.

EXAMPLE 1—

A plane leaves Chicago going west. A faster plane going 40 mph faster leaves Chicago at the same time going east. After two hours the planes are 2000 miles apart. Find the speeds of the planes.

I draw what I call my crummy little diagram, to see what is happening.

	2x	Chicago	2(x + 40)	
		•		
		— 2000 —		

	Rate	×	time	=	distance
West	x		2		2x
East	x + 40		2		2(x + 40)

From the picture, the two distances add to 2000 miles.

$2x + 2(x + 40) = 2000$

$4x + 80 = 2000$

$4x = 1920$

$x = 480$ miles/hour

$x + 40 = 520$ miles/hour

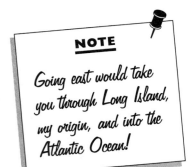

NOTE

Going east would take you through Long Island, my origin, and into the Atlantic Ocean!

EXAMPLE 2—

A train leaves New York City going west. An express going 20 mph faster leaves New York City 2 hours later on the same track going west. 6 hours later the express hits the slower train. Find their speeds when the express hits the slower train.

Rate	×	time	=	distance
Slower	x		8	8x
Express	x + 20		6	6(x + 20)

The express goes for 6 hours. The slower train goes for 2 + 6 = 8 hours. If x = speed of the slower train, x + 20 = speed of the faster one. They start at the same point and wind up in the same spot when they crash (even though the times are different). From the picture we can see that the distances are equal.

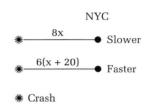

8x = 6(x + 20)

2x = 120

x = 60

x + 20 = 80

The slower train was going 60 mph and the express 80 mph. You'll all be happy to know that no one was hurt.

EXAMPLE 3

A car makes a trip going 40 mph. The return trip is made at 60 mph. What is the average speed?

This is the kind of problem that the SAT asks. First, notice that the distance is not given. The easiest way to do the problem is to assign a distance. In this case, we choose 120 miles, since it is the smallest number that both 40 and 60 divide into with no remainder.

Next, we fill out the chart.

	Rate	×	time	=	distance
First trip	60				120
Return trip	40				120

Since the time = distance divided by the rate, the time of the first trip is 120/60 = 2 hours.

The return trip time = 120/40 = 3 hours.

The average rate is the total distance/total time = 240/5 = 48 mph!

The average rate is not, *not, NOT* the average of the rates!!!!

You can find more about SAT problems in *SAT® Math for the Clueless.*

Let's move on to a very important chapter.

CHAPTER 5

FACTORING

This is one of my favorite chapters. Factoring is a game. I really like games, especially this one. First we need to learn a little bit more about the distributive law. Then we'll do a lot of factoring, followed by a few applications.

MORE ABOUT THE DISTRIBUTIVE LAW

We have been doing problems like multiplying $3x(5x^2 - 7x + 5y - 9) = 15x^3 - 21x^2 + 15xy - 27x$. We have been multiplying monomials times polynomials. We would like to do other multiplications.

First, we would like to multiply a binomial times a binomial using the FOIL method. FOIL is an *acronym*. F = first (in each parenthesis), O = outer, I = inner, and L = last. Let's do a few.

An *acronym* is a word where each letter stands for a word. Three of my favorites are SCUBA—self-contained underwater breathing apparatus, NATO—North Atlantic Treaty Organization, and an old army acronym, SNAFU—situation normal, all fouled up!

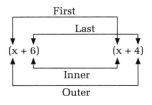

First

Last

(x + 6)   (x + 4)

Inner

Outer

EXAMPLE I—

Multiply $(x + 6)(x + 4)$.

First $(x)(x) = x^2$, outer $(x)4 = 4x$, inner $(6)x = 6x$, last $(+6)(+4) = +24$. Add inner and outer and we get $4x + 6x = 10x$. Final answer: $x^2 + 10x + 24$.

It is important to be able to do these problems very quickly. You want to be able to write ONLY the answer. Terms are written from the highest exponent to lowest and alphabetically if there are two letters.

EXAMPLE 2—

Multiply $(3x - 5y)(7x + 2y)$.

first $3x(7x) = 21x^2$, outer $(3x)(2y) = 6xy$, inner $(-5y)(7x) = -35xy$, inner + outer $= -29xy$, last $(-5y)(2y) = -10y^2$. Answer: $21x^2 - 29xy - 10y^2$.

EXAMPLE 3—

$(3x - 5)^2$. Squaring is also foiling.

$(3x - 5)^2 = (3x - 5)(3x - 5) = 9x^2 - 15x - 15x + 25$

$\qquad = 9x^2 - 30x + 25$

You should try to square in your head not only for this chapter but for later. It will make other sections easier and faster to learn.

In words:

$(a + b)^2 = a^2 + 2ab + b^2$

$(\text{first} + \text{second})^2 = \text{first}^2 + 2 \times \text{first} \times \text{second} + \text{second}^2$

$(a - b)^2 = a^2 - 2ab + b^2$

$(\text{first} - \text{second})^2 = \text{first}^2 - 2 \times \text{first} \times \text{second} + \text{second}^2$

EXAMPLE 4—

$(3ab + c^3)(3ab - c^3) = 9a^2b^2 - 3abc^3 + 3abc^3 - c^6 = 9a^2b^2 - c^6$

Notice that the middle terms kill each other.

$(a + b)(a - b) = a^2 - b^2$

In words: (first + second)(first − second) = first² − second², *the difference of two squares.*

The first parentheses can have the minus sign and the second the plus sign, and the answer will be the same (the commutative law says so).

To finish this lesson, we want to multiply any polynomial by another polynomial. While it is very important to be able to do the last part in your head, it is not as important here. Let us multiply a trinomial by a trinomial. We will show two methods.

EXAMPLE 5—

$(2x^2 + \ 3x \ + 4)(3x^2 + 4x \ + 5)$

$2x^2 + \ 3x \ + 4$

$* \ 3x^2 + \ 4x \ + 5$

$\overline{}$

$6x^4 + \ 9x^3 + 12x^2$ **Multiply $3x^2 \times (2x^2 + 4x + 4)$.**

$\qquad \quad 8x^3 + 12x^2 + 16x$ **Multiply $4x \times (2x^2 + 3x + 4)$, lining up like terms.**

$\qquad \qquad \qquad 10x^2 + 15x + 20$ **Multiply $5 \times (2x^2 + 3x + 4)$.**

$\overline{}$

$6x^4 + 17x^3 + 34x^2 + 31x + 20$ **Add like terms.**

Try to do this in your head. No, no, it is not that bad. Many of you will be able to do it. It will sharpen your basic algebra a lot!!!!

$(2x^2 + 3x + 4)(3x^2 + 4x + 5)$

Multiply the two highest powers:

$(2x^2)(3x^2) = 6x^4$

Look for x^3 terms:

$(2x^2 + 3x + 4)(3x^2 + 4x + 5)$

$2x^2(4x) = 8x^3$ and $(3x)(3x^2) = 9x^3$

$8x^3 + 9x^3 = 17x^3$

Look for x^2 terms:

$(2x^2 + 3x + 4)(3x^2 + 4x + 5)$

$(2x^2)(5) = 10x^2$ and $(3x)(4x) = 12x^2$ and $(4)(3x^2) = 12x^2$

$10x^2 + 12x^2 + 12x^2 = 34x^2$

Look for x terms:

$(2x^2 + 3x + 4)(3x^2 + 4x + 5)$

$(3x)(5) = 15x$ and $(4)(4x) = 16x$

$15x + 16x = 31x$

Lastly:

$(4)(5) = 20$

ANSWER
$6x^4 + 17x^3 + 34x^2 + 31x + 20$

It is a lot easier to do in your head than to write out.
Try it. You may like it . . . a lot.

EXAMPLE 6—
Multiply $(3x^2 - 5)(4x^3 - 5x - 7)$. Notice that the exponents are *not* consecutive.

$$4x^3 - 5x - 7$$
$$\underline{\times \quad 3x^2 - 5}$$
$$12x^5 - 15x^3 - 21x^2$$

Multiply $3x^2 \times 4x^3 - 5x - 7$; leave space for missing exponents—sometimes they are filled in and sometimes not.

$$-20x^3 \qquad + 25x + 35$$

$-5 \times 4x^3 - 5x - 7$; line up like terms.

$$12x^5 - 35x^3 - 21x^2 + 25x + 35$$

Add like terms.

ANSWER

$12x^5 - 35x^3 - 21x^2 + 25x + 35$

Try to do this one in your head also.

THE GAME OF FACTORING

This is one of my favorite sections. It is a puzzle or a game, depending on how you look at it. There are various types. Let's learn the rules so we can play.

The Largest Common Factor

We learned that if we multiply $3(2x - 5)$, we get $6x - 15$. Given $6x - 15$, we would like to arrive at $3(2x - 5)$. In other words, factoring is the distributive law backward.

EXAMPLE 1—

Factor completely $8x - 12y$.

We ask, What is the largest factor multiplying both $8x$ and $-12y$? The answer is 4. 4 times what is $8x$? The answer is $2x$. 4 times what is $-12y$? $-3y$. So, $8x - 12y$ factors into $4(2x - 3y)$.

EXAMPLE 2—

Factor completely $10x^2 - 15xy + 25xz$.

$5x$ is the common factor of each term: $10x^2 - 15xy + 25xz = 5x(2x - 3y + 5z)$.

If you want to check, just multiply $5x(2x - 3y + 5z)$ and get $10x^2 - 15xy + 25xz$.

EXAMPLE 3—

Factor completely $27x^6y^7z^8 + 36x^{80}y^2z^{11}w^5 + 18x^{1000}y^5zw^4$.

9 is the common number factor. The lowest power of each letter can be factored out (that would be the largest number of factors *in common*): x^6, y^2, z (lowest exponent is 1), no w's because w is not in each term! Soooo,

$27x^6y^7z^8 + 36x^{80}y^2z^{11}w^5 + 18x^{1000}y^5zw^4$

$= 9x^6y^2z(3y^5z^7 + 4w^5x^{74}z^{10} + 2w^4x^{994}y^3)$

NOTES

1. To know we have taken out the largest number, in the parentheses, the largest common factor of the numbers 3, 4, and 2 must be 1.

2. In the parentheses, there is no common letter in each of the terms: w and x in the second and third terms, but not the first; z in the first and second, but not the third; y in the first and third, but not the second.

3. Of course, you can multiply it out to make the final check.

4. Large numbers do *not* make hard problems.

5. Practice these until you are very good at them, and please don't forget. Most students learn this well and then forget. You must always take the largest common factor out first.

EXAMPLE 4—

Factor completely 6ab + 8ac + 9bc.

The answer is that this expression does not factor (it is called *prime*).

2 is a factor of 6 and 8, but not 9. 3 is a factor of 6 and 9, but not 8.

a is a factor of the first and second, but not the third. b is a factor of the first and third, but not the second.

c is a factor of the second and third, but not the first. No common factor: *prime*.

EXAMPLE 5—

Factor $ab^3 + a^2b + ab$.

The common factor is ab. $ab^3 + a^2b + ab = ab(b^2 + a + 1)$.

Note that if there are three terms at the start, then three terms must be in the parentheses. If *everything* is factored out, we must put a 1 because 1(ab) = ab.

We can now finish our lesson on linear equations. Remember, there was a step 7 missing. Here it is:

7. Factor out an x (common factor). This only occurs if there are at least two letters in the problem.

EXAMPLE 6—

Solve for x:

$$y = \frac{3x - 7}{2x + 9}$$

Cross multiply to clear fractions (or multiply each side by LCD, 2x + 9).

$$\frac{y}{1} = \frac{3x - 7}{2x + 9}$$

Distributive.

$$3x - 7 = y(2x + 9)$$

$$3x - 7 = 2xy + 9y$$

Get all the x terms to the left; get the non-x terms to the right. When they switch sides, they change signs. (Hope you noticed.)

Factor out the x (step 7).

$$3x - 2xy = 9y + 7$$

$$x(3 - 2y) = 9y + 7$$

Divide each side by the coefficient of x, which is 3 − 2y.

$$\frac{x(3 - 2y)}{3 - 2y} = \frac{9y + 7}{3 - 2y}$$

$$x = \frac{9y + 7}{3 - 2y}$$

EXAMPLE 7—

Solve for p: $A = p + prt$.

The formula says the amount of money, A, equals the principle + the simple interest (principle times rate times time).

Put the variable you are solving for on the left.

$$A = p + prt$$

Factor out the p.

$$p + prt = A$$

$$p(1 + rt) = A$$

$$p = \frac{A}{1 + rt}$$

Divide by the whole coefficient of p.

In school, teachers used to give 15 or 20 of this type. Now few teachers give any, but they are needed for algebra and have been found on the SAT.

Difference of Two Squares

As the name says, we have two square terms with a minus sign between them:

$(\text{first})^2 - (\text{second})^2 = (\text{first} + \text{second})(\text{first} - \text{second})$

or $(\text{first} - \text{second})(\text{first} + \text{second})$

EXAMPLE 1—

Factor completely $x^2 - 16$.

$(x + 4)(x - 4)$

EXAMPLE 2—

Factor completely $16c^2 - 25d^2$.

$(4c - 5d)(4c + 5d)$

EXAMPLE 3—

Factor completely $(\text{pig})^2 - (\text{cow})^2$.

$(\text{pig} + \text{cow})(\text{pig} - \text{cow})$

EXAMPLE 4—

Factor completely $c^8 - m^{10}$.

$(c^4 - m^5)(c^4 + m^5)$

Remember, $m^5 m^5 = m^{10}$.

EXAMPLE 5—

Factor completely $a^2 + b^2$.

This does not factor; the sum of two squares doesn't factor. *Try!!!!*

EXAMPLE 6—

Factor completely $(a + b)^2 - 64$.

$[(a + b) + 8][(a + b) - 8]$

Sometimes there is more than one factoring in a problem. First is the largest common factor; then the difference of two squares.

EXAMPLE 7—

Factor completely $3x^2 - 27$.

The common factor is 3: $3(x^2 - 9)$. Then the difference of two squares. So $3x^2 - 27 = 3(x - 3)(x + 3)$.

EXAMPLE 8—

Factor completely $x^4 - 16$.

No common factor. Difference of two squares: $x^4 - 16 = (x^2 + 4)(x^2 - 4)$. $x^2 + 4$ is the sum of two squares; no more factoring, but $x^2 - 4$ factors. So, for the complete factorization:

$$x^4 - 16 = (x^2 + 4)(x^2 - 4) = (x^2 + 4)(x + 2)(x - 2)$$

Most of my students find this is the easiest factoring, requiring the least amount of practice. But everyone is an individual. Some find easy sections harder and harder sections easier. The words *easy* and *hard* are for large groups. You are not a group, but an important individual.

Factoring Trinomials I

This section is much easier to demonstrate on the board than it is to write. If you don't like this, write me a nasty letter (well, not too nasty), and tell me what you don't understand. You might also write what you like.

In the section "More About the Distributive Law," we multiplied out the following:

first ⎯⎯⎯⎯⎤ ⎡⎯⎯⎯ last

$(x + 6)(x + 4) = x^2 + 10x + 24$

$(x - 7)(x - 2) = x^2 - 9x + 14$

$(x + 5)(x - 3) = x^2 + 2x - 15$

$(x + 5)(x - 8) = x^2 - 3x - 40$

We would like to go backward.

Since the first term is always x^2 (coefficient +1), we call the first sign and last sign as indicated by the arrows.

Rules of signs for factoring trinomials:

Statement	Reason
1. If the last sign in the trinomial is plus, both of the signs in the two factors are the same. (first two problems).	1. Plus times a plus is a plus: $(+6)(+4) = +24$. Minus times minus is a plus: $(-7)(-2) = +14$.
2. *Only if the last sign is plus,* look at the first sign.	2. (plus) + (plus) = plus: $+4x + +6x = +10x$.
First sign of trinomial plus, both signs in the factors plus. First sign of the trinomial minus, both signs of the factors minus.	(minus) + (minus) = minus: $(-2x) + (-7x) = -9x$.
3. If the last sign in the trinomial is minus, the factors have one plus sign and one minus sign.	3. Plus times a minus is a minus.

We'll see which goes where as we do the problems. Now that we know the signs, we'll see how the game is played.

EXAMPLE 1

Factor completely $x^2 + 7x + 6$.

Last sign is plus (+6), so each of the factors have the same sign. First sign plus (+7x) means both signs of the factors are plus. We look at the first term, x^2. x^2 always equals (x)(x). So far we have $x^2 + 7x + 6 =$ (x +)(x +).

NOTE

The game gets more complicated if the first term is $2x^2$ or $6x^2$.

We then look at last term and find all factors of 6: 6 = (2)(3) or (1)(6). Let us try them:

$(x + 2)(x + 3) = x^2 + 5x + 6$ Wrong!!!!

$(x + 1)(x + 6) = x^2 + 7x + 6$ Correct!!!!

NOTES

1. This is why you need to know the multiplications verrrry well.

2. $(x + 6)(x + 1)$ is also right.

Not too bad. Let's try some more.

EXAMPLE 2

Factor completely $c^2 - 8c + 12$.

Last sign plus, first sign minus means both factors have minus signs.

$c^2 = c(c)$ $12 = 1(12) = 6(2) = 4(3)$

2 and 6 check.

ANSWER
$(c - 2)(c - 6)$

EXAMPLE 3—

Factor completely $z^2 - 6z - 27$.

Last sign negative means one factor plus and one factor minus. But which one? Just wait a minute; you'll see.

$$z^2 = (z)(z) \qquad 27 = 9(3) = 27(1)$$

9 and 3 work, but which has the plus sign? Let's try $(z + 9)(z - 3)$. If we multiply it out, we get $z^2 + 6z - 27$. If the middle number is right but the *sign* is wrong, we must change *both signs* of the factors.

ANSWER

$$z^2 - 6z - 27 = (z - 9)(z + 3)$$

EXAMPLE 4—

Factor completely $a^2 + 7ab - 8b^2$.

Last sign minus, factors have different signs.

8b and 1b = b are correct; minus with the b.

$$8b^2 = (4b)(2b) = (8b)(1b)$$

ANSWER

$(a - b)(a + 8b)$

EXAMPLE 5—

Factor completely $b^2 + 6b + 10$.

Factors both have plus signs. 10 = 5(2) or 10(1), but neither checks. Since these are the only possibilities, $b^2 + 6b + 10$ is *prime.*

Sometimes, there are two or more factorings.

EXAMPLE 6—

Factor completely $3cd^2 - 12cd + 9c$.

Common factor is 3c: $3cd^2 - 12cd + 9c = 3c(d^2 - 4d + 3)$.

No difference of two squares.

$d^2 - 4d + 3$: last sign plus, first sign minus means both factors are minus; $3 = 3(1)$. Soooo,

$3cd^2 - 12cd + 9c = 3c(d^2 - 4d + 3) = 3c(d - 3)(d - 1)$

The order is:

1. Largest common factor (almost always first if it occurs).

2. Difference of two squares (almost always second, if it occurs).

4. Trinomial (almost always fourth, if it occurs).

Factoring Trinomials II

The game gets a little more interesting now. I like the challenge.

EXAMPLE 1

Factor completely $2m^2 + 7m + 3$.

Last sign and first sign both plus means both factors have a plus sign. $2m^2 = 2m(m)$ (only combination) and $3 = 1(3)$ (only combination). But because the coefficents of the factors are different, we must locate where the 1 and 3 go. (Before, with a plus sign, we didn't have to worry.)

$(2m + 3)(m + 1) = 2m^2 + 5m + 3$ Wrong!!!!

$(2m + 1)(m + 3) = 2m^2 + 7m + 3$ Right!!!!!

EXAMPLE 2

Factor completely $3x^2 - 7xy + 4y^2$.

Last sign plus, first sign minus means both factors are minus.

$3x^2 = 3x(x)$ $4y^2 = 2y(2y) = 4y(y)$

2y and 2y you don't have to reverse, but you must reverse 4y and y.

$(3x - 2y)(x - 2y) = 3x^2 - 8xy + 4y^2$ Wrong!!!

$(3x - y)(x - 4y) = 3x^2 - 13xy + 4y^2$ Wrong!!!!

$(3x - 4y)(x - y) = 3x^2 - 7xy + 4y^2$ Right!!!!!

Getting a little more challenging, isn't it??!!!!

EXAMPLE 3—

Factor completely $9x^2 - 8 - 21x$.

Problems like this can be very long. Here are the steps:

Step 1.   Arrange the terms from highest exponent to lowest: $9x^2 - 21x - 8$. If there are two letters, one letter should be highest exponent to lowest.

Step 2.   Determine all the signs. Last sign negative means both factors have different signs. Put in the signs at the end unless they are the same.

Step 3.   Write all factors of 8: $8 = 2(4)$ or $8(1)$.

Step 4.   Write out all possibilities: 3x and 3x reversing is not necessary; 9x and 1x reversing is necessary, but do the reversing with the numbers.

1. $(9x \quad 4)(x \quad 2)$

2. $(9x \quad 2)(x \quad 4)$

3. $(9x \quad 8)(x \quad 1)$

4. $(9x \quad 1)(x \quad 8)$

5. $(3x \quad 4)(3x \quad 2)$

6. $(3x \quad 1)(3x \quad 8)$

There are six possibilities.

Step 5.   Since we picked first terms to be correct and last terms to be correct, we only have to find inner and outer.

1. 4x and 18x, close is only good in horseshoes and hand grenades

2. 2x and 36x, no good

3. 8x and 9x, no good

4. 1x and 72x, the next building

5. 12x and 6x, no good

6. 3x and 24x, at last correct; fill in the correct signs!

ANSWER
$(3x + 1)(3x - 8)$

I made sure the correct answer was last just to show you all the possibilities. As soon as you find the correct combination, you stop. The more you practice, the quicker you will see many of these.

EXAMPLE 4—

Factor completely $4x^3 - 12x^2 + 9x$.

Again, sometimes there are multiple factors.

Common factor is x: $4x^3 - 12x^2 + 9x = x(4x^2 - 12x + 9)$.

$4x^2 - 12x + 9$: both factors are minus; $4x^2 = 2x(2x)$ or $x(4x)$; $9 = 3(3)$ or $9(1)$.

The correct one is:

$4x^3 - 12x^2 + 9x = x(2x - 3)(2x - 3) = x(2x - 3)^2$

NOTE
When doing these problems, try the x terms that are closest together first. The way books are written and most teachers give problems, the closer ones are correct more often.

EXAMPLE 5

If $\left(x + \dfrac{1}{x}\right)^2 = 100$, what is $x^2 + \dfrac{1}{x^2}$?

This is the SAT type question I call a "phony phactoring" or a "fony factoring."

Actually, the trick is just to square the binomial!

$$\left(x + \frac{1}{x}\right)^2 = x^2 + 2(x)\left(\frac{1}{x}\right) + \frac{1}{x^2} = x^2 + 2 + \frac{1}{x^2} = 100$$

Subtracting 2 from each side, we get:

$$x^2 + \frac{1}{x^2} = 98!$$

Sum and Difference of Cubes

Although only the difference of two squares factors, both the sum and difference of cubes factor.

$x^3 - y^3 = (x - y)(x^2 + xy + y^2)$

$(\text{first})^3 - (\text{second})^3 = (\text{first} - \text{second})[(\text{first})^2 + (\text{first})(\text{second}) + (\text{second})^2]$

aaaand

$x^3 + y^3 = (x + y)(x^2 - xy + y^2)$

$(\text{first})^3 + (\text{second})^3 = (\text{first} + \text{second})[(\text{first})^2 - (\text{first})(\text{second}) + (\text{second})^2]$

You should check by multiplying both out.

EXAMPLE 1—

Factor completely $8x^3 - 27y^3$.

$8x^3 - 27y^3 = (2x)^3 - (3y)^3$. First = 2x, second = 3y.

$8x^3 - 27y^3 = (2x - 3y)(4x^2 + 6xy + 9y^2)$

EXAMPLE 2—

Factor completely

$125a^6 + 64b^9 = (5a^2)^3 + (4b^3)^3 = (5a^2 + 4b^3)$

$\times (25a^4 - 20a^2b^3 + 16b^6)$

EXAMPLE 3—

Factor completely $x^6 - y^6$.

This is a double factoring. x^6 and y^6 are both squares and cubes. But squares come before cubes.

$x^6 - y^6 = (x^3 - y^3)(x^3 + y^3) = (x - y)(x^2 + xy + y^2)(x + y)$

$\times (x^2 - xy + y^2)$

Terms in any order.

The order of factoring is:

1. Greatest common factor (any number of terms)

2. Difference of two squares (two terms)

3. Sum and difference of two cubes (two terms)

4. Trinomial (three terms)

The order is almost always absolute . . . until we get to the last factoring, grouping (more than three terms).

Grouping

EXAMPLE 1—

Factor completely $ax + ay + bx + by$.

There is no 1, 2, 3, or 4 buuuut if we look at ax + ay + bx + by, from the first two terms, there is a common factor of a and from the last two terms, there is a common factor of b. Soooo,

ax + ay + bx + by = a(x + y) + b(x + y)

Now the common factor is x + y.

So the answer is:

ax + ay + bx + by = a(x + y) + b(x + y) = (x + y)(a + b)

EXAMPLE 2—

$c^2 - 6c + 9 - d^2$

Not only can the factoring be two and two but three and one (or one and three).

$c^2 - 6c + 9 - d^2 = (c - 3)(c - 3) - d^2 = (c - 3)^2 - d^2$

$= (c - 3 + d)(c - 3 - d)$

Difference of two squares.

EXAMPLE 3—

$ax + ay + x^2 + 6xy + 5y^2$

First two terms common factor; Last three trinomial.

$ax + ay + x^2 + 6xy + 5y^2 = a(x + y) + (x + 5y)(x + y)$

$= (x + y)(a + x + 5y)$

They can get longer. I love puzzles like these. I hope you will too.

Here's a challenger. Factor: $x^4 + x^2 + 1$. The answer is at the end of the chapter. (Yes, it does factor!)

We are now ready for applications. Factoring is part of algebra, trigonometry, and calculus, an important part. For the rest of this chapter and a large part of the next one, you will get a lot more practice.

SOLVING POLYNOMIAL EQUATIONS BY FACTORING

Factors Whose Product Is Zero

If you multiply two numbers and the product is zero, what can you say about the numbers?

Symbolically, if ab = 0, then either a = 0 or b = 0. Suppose a = 0. Then ab = 0(b) = 0. Suppose a ≠ 0 and ab = 0. We can divide both sides by a, annnnd

$$\frac{ab}{a} = \frac{0}{a} \quad \text{or} \quad b = 0$$

Yes, this is a proof I sneaked in. Not all proofs are hard!!!!

EXAMPLE 1—

Solve for all values of x:

(x − 5)(x + 6) = 0

Using this property . . . x − 5 = 0 or x + 6 = 0. So the roots are x = 5 or −6.

EXAMPLE 2—

Solve for all values of x:

x(x + 4)(x − 5)(2x − 7)(5x + 1) = 0

Set each factor equal to zero. We get x = 0 or x + 4 = 0 or x − 5 = 0 or 2x − 7 = 0 or 5x + 1 = 0.

The roots are 0, −4, 5, 7/2, −1/5.

It would be very useful to be able to solve this part of the equation in your head. If ax + b = 0, solving for x, x = −b/a. In words, the opposite sign of b divided by the coefficient of x.)

Solving Equations by Factoring

We are going to solve *polynomial equations,* equations where the left and right sides are polynomials, by factoring. *Quadratic equations,* or *second-degree equations,* where the highest exponent is 2, have two solutions. *Cubic* or *third-degree equations,* where the highest power is 3, have three roots. *Quartic* or *fourth-degree equations* have four answers, and so on.

The beginning of solving by factoring is similar to solving linear equations. But let us list the steps, just like before.

Solving equations by factoring

1. Multiply all terms by LCD to get rid of all fractions.

2. Multiply out all ( ) and [ ].

3. Simplify each side by combining like terms.

4. Add opposites of all the terms on the right to each side, writing the terms highest exponent to lowest exponent. The right side at this point should be 0.

5. If the coefficient of the highest power is negative, multiply each term by −1. (Easier to factor.)

6. Factor.

7. Set each factor equal to 0 and write the roots.

Okay, let's do some. If your factoring is good, this section is very easy.

EXAMPLE I—

Solve for all values of x:

$x^2 - 4x - 21 = 0$

The equation is already on the left side, highest power to lowest. (This is called *standard form*.) All we need do is factor:

$(x - 7)(x + 3) = 0$

The roots are 7 and −3.

EXAMPLE 2—

Solve for all values of x:

$4x^2 + 16x + 16 = 0$

Again, this is in standard form. Factoring, we get:

$4(x^2 + 4x + 4) = 4(x + 2)(x + 2) = 0$

$x + 2 = 0 \qquad \text{or} \qquad x + 2 = 0$

In this case we get two roots, but they are equal. Soooo, the solution is $x = -2$.

EXAMPLE 3—

Solve for all values of y:

$y^2 + 24 = 10y$

We must get all the y's to the left.

$-10y = -10y$

Terms must be highest to lowest.

$y^2 - 10y + 24 = 0$

Factoring, $(y - 4)(y - 6) = 0$. The roots are 4 and 6.

EXAMPLE 4—

Solve for all values of z:

$(z - 3)(z - 4) = 6$

Multiply out ( ).

$z^2 - 7z + 12 = 6$

Add −6 to both sides.

$-6 = -6$

Factor.

$z^2 - 7z + 6 = 0$

$(z - 6)(z - 1) = 0$

So $z = 6$ or 1.

EXAMPLE 5—

Solve for all values of b:

$$-b - \frac{8}{b} = -9$$

$$\left(\frac{b}{1}\right)\frac{(-b)}{1} - \frac{b}{1} \times \frac{8}{b} = \frac{-9}{1} \times \frac{b}{1}$$

Multiply every term by $\frac{b}{1}$, the LCD.

$$-b^2 - 8 = -9b$$

$$+9b = +9b$$

Add 9b to both sides.

$$-b^2 + 9b - 8 = 0$$

Multiply every term by −1.

$$b^2 - 9b + 8 = 0$$

$$(b - 8)(b - 1) = 0$$

Factor.

So b = 8 and 1.

NOTE

If one of the answers turned out to be b = 0, it would not be allowed because we are not allowed to divide by 0. If an equation had a term $7/(x - 5)$, x = 5 could not be a root.

EXAMPLE 6—

Solve for all values of a:

$$a^3 + 12a^2 - 13a = 0$$

This is a cubic, so it has three solutions.

$$a^3 + 12a^2 - 13a = a(a^2 + 12a - 13) = a(a + 13)(a - 1) = 0$$

The roots are 0, −13, and 1.

Some of you sharp-thinking students might ask, "What if the equation does not factor?" It is a very good question. Later we will learn to solve all quadratic equations. We will solve some cubics, quartics, and fifth-degree equations much later. It's not too hard.

Word Problems

Some word problems are solved by factoring. We will do some here and some later.

EXAMPLE 1—

If a square has its base doubled and its height increased by 3 feet, the area is increased by 40 square feet. Find the length of the original side of the square.

Draw a picture (actually, two pictures). Let x = side of the square.

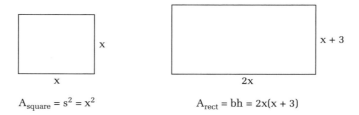

$A_{square} = s^2 = x^2$ $\qquad$ $A_{rect} = bh = 2x(x + 3)$

Area of the square increased by 40 = area of the rectangle

$x^2 + 40 = 2x(x + 3)$

$x^2 + 40 = 2x^2 + 6x$

$-2x^2 - 6x = -2x^2 - 6x$

Multiply both sides by −1. $\quad -x^2 - 6x + 40 = 0$

$x^2 + 6x - 40 = 0$

$(x + 10)(x - 4) = 0$

So x = −10 and x = 4. x = −10 cannot happen. A square can't have a length of −10 feet, at least not on this planet. The square has a side of 4 feet.

Not too bad. . . . Most students like most of these problems more than those in the last chapter. One problem I
like to deal with in great detail is the next one because
it is found later in math, physics, and other areas.

EXAMPLE 2—

An object is thrown upward. Its equation is given by
height $y = -16t^2 + 64t + 80$, y in feet, t = time in seconds.

   a. Find the height after 2 seconds.

   b. When is the object 128 ft high?

   c. When does it hit the ground?

ANSWERS

   a. Easy: t = 2

$$y = -16t^2 + 64t + 80 = -16(2)^2 + 64(2) + 80 = 144 \text{ ft}$$

   b. y = 128

$$-16t^2 + 64t + 80 = 128$$

$$-16t^2 + 64t - 48 = 0$$

$$-16(t^2 - 4t + 3) = 0$$

$$-16(t - 1)(t - 3) = 0$$

Soooo, t = 1 or t = 3. Why two answers? Onnce
on the way up, and once on the way down.

   c. The ground means y = 0.

$$0 = -16t^2 + 64t + 80 = 0$$

so

$$16t^2 - 64t - 80 = 0 \quad \text{or} \quad 16(t - 5)(t + 1) = 0$$

Time t = 5 seconds or −1 seconds. We cannot
have negative time because we start with time

$t = 0$. The answer is 5 seconds. Notice that this formula only holds until 5 seconds. When the object hits the ground, the formula changes. If it's a ball, it will bounce, but not as high; if it's a spear, it goes in the ground; if it's an egg, it goes splat. But the formula changes.

Let's talk more about the problem. The general formula is

$$y = -16t^2 + v_0t + y_0$$

where

y = height in feet

t = time in seconds

v_0 = velocity at time $t = 0$ (Speed—but the difference is, speed is always positive; velocity can be negative.)

v_0 = feet/second (v_0 is positive if the object is thrown upward, negative if it is thrown down, and 0 if it is just dropped.)

y_0 = height at time $t = 0$ (y_0 can be positive if you start, let's say, from the top of a building. $y_0 = 0$ from the ground. y_0 can't be negative. It is very difficult to throw something upward from under the ground. The dirt gets in the way.)

NOTE

The zeroes in v_0 and y_0 are called *subscripts*. They are tags (labels). They are *not* exponents.

Let's examine the formula.

$-16t^2$ comes from the formula $s = \frac{1}{2}at^2$. a = acceleration = gravity = -32 ft/sec². The minus sign indicates that gravity is down. It would be verry interesting if gravity were up.

v_0t we have already studied: rate × time = distance.

So the last problem could have been the following picture:

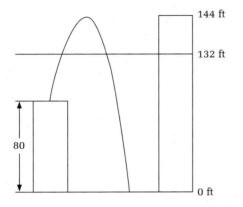

At this point we can't show that the highest point is 144 ft, but it is. Later we will prove it, we will!!!!*

We should also know this formula in meters: $y = -4.9t^2 + v_0t + y_0$ in meters, because gravity is -9.8 meters/second^2.

The next chapter also has a lot of factoring. Most of it is also very, very important.

EXAMPLE 3—

In a right triangle, the sides are consecutive integers. Find them.

We let x and x + 1 be the legs. The hypotenuse, the largest side, must be x + 2.

We get $x^2 + (x + 1)^2 = (x + 2)^2$.

Multiplying and simplifying, we get $2x^2 + 2x + 1 = x^2 + 4x + 4$ or $x^2 - 2x - 3 = 0$.

Factoring, we get $(x - 3)(x + 1) = 0$ or x = 3 or −1. x cannot be −1. The length of a side of a triangle can't be

*In *Precalc with Trig for the Clueless*.

negative, in the same way that you can't be minus six feet tall. Maybe in some other universe, but not in ours. So x is 3, and the sides are 3, 4, and 5.

Answer to the challenger: If the problem was $a^2 + a + 1$, it could not factor.

The only factors of $x^4 = x^2 \times x^2$. The only factors of 1 are 1×1.

If we try $(x^2 + 1)^2 = (x^2 + 1)(x^2 + 1) = x^4 + 2x^2 + 1$, an extra x^2. So we subtract the extra x^2.

We now have $(x^2 + 1)^2 - x^2$. But if we look carefully, this becomes the difference of 2 squares.

$(x^2 + 1)^2 - x^2 = (x^2 + 1 + x)(x^2 + 1 - x)$!!!!

Multiply it out to check it. Factoring can be quite a game!

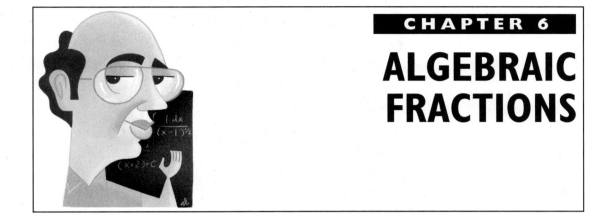

CHAPTER 6

ALGEBRAIC FRACTIONS

At the beginning, we are going to do the same as we do with number fractions: reduce (or simplify) them, multiply and divide them, and add and subtract them. You might want to review the section on fractions in Appendix 1.

REDUCING

To reduce a fraction, we must factor the top, factor the bottom (if not already factored) and cancel the common factor(s). In symbols:

$$\frac{mk}{nk} = \frac{m}{n} \qquad k \neq 0$$

Remember also $n \neq 0$, the denominator can never be zero.

EXAMPLE 1—

Reduce

The bottom can nevvvver = 0!

$$\frac{10a + 15}{8a^2 + 10a - 3} = \frac{5(2a + 3)}{(2a + 3)(4a - 1)} = \frac{5}{4a - 1}$$

$$a \neq \frac{-3}{2} \text{ or } \frac{1}{4}$$

EXAMPLE 2—

Reduce

$$\frac{4 - b}{b^2 - 16} = \frac{4 - b}{(b - 4)(b + 4)} = \frac{-1(b - 4)}{(b - 4)(b + 4)} = \frac{-1}{b + 4}$$

$$b \neq 4 \text{ or } -4$$

$$\frac{b - 4}{4 - b} = -1$$

In general,

$$\frac{a - b}{b - a} = -1$$

Top is exactly the negative of the bottom!!!!

$$\frac{a - b - c + d - e}{-a + b + c - d + e} = -1$$

EXAMPLE 3—

Reduce

$$\frac{4c^6 d^7 e^8}{6c^2 d^7 e^{10}}$$

Remember, this is also reducing.

ANSWER

$$\frac{2c^4}{3e^2} \qquad c, d, e \neq 0$$

But exactly when can we reduce and when can't we reduce????

Yes:

$$\frac{4(3)}{3} = \frac{4}{1} = 4$$

If we have letters?

$$\frac{b(c)}{c} = \frac{b}{1} = b \qquad c \neq 0$$

We can reduce because there is a common factor top and bottom.

No:

$$\frac{4 + 3}{4} = \frac{7}{4} \neq 3$$

Can't cancel 4s because of order of operation.

If we have letters?

$$\frac{b + c}{b} = \frac{b + c}{b} \qquad b \neq 0$$

Can't cancel the b's for the same reason.

We can't reduce because top and bottom have no common factor (except 1 or −1).

MULTIPLICATION AND DIVISION

The rules are the same as for number fractions.

Multiplication $\qquad \left(\dfrac{a}{b}\right)\left(\dfrac{c}{d}\right) = \dfrac{ac}{bd}$

Division $\qquad \dfrac{e}{f} \div \dfrac{g}{h} = \left(\dfrac{e}{f}\right)\left(\dfrac{h}{g}\right) = \dfrac{eh}{fg}$

Remember, before you multiply or divide, you must factor all the tops and bottoms (unless they are already factored), cancel one factor in any top with one factor in any bottom (in division, after you invert the second fraction), and then multiply the tops and multiply the bottoms.

EXAMPLE 1—

$$\frac{8x^2y^5}{6a^9b^{10}} \times \frac{9x^5a^4}{10b^6y^2} = \frac{\overset{2}{\cancel{8}}\ x^2\overset{y^3}{\cancel{y^5}}}{\underset{1}{\cancel{6}}\underset{\cancel{a^5}}{\cancel{a^9}}b^{10}} \frac{\overset{3}{\cancel{9}}x^5\overset{\cancel{a^4}}{}}{10b^6\cancel{y^2}} = \frac{6x^7y^3}{5a^5b^{16}} \qquad a, b, y \neq 0$$

EXAMPLE 2—

Factor everything!! You get lots of practice!!!

$$\frac{x^2 - 9}{x^2 - 9x} \times \frac{x^6}{x^2 + 6x + 9} \div \frac{2x^2 + 7x + 3}{4x^2 - 1}$$

Invert. Note that steps 2 and 3 can be interchanged.

$$= \frac{(x + 3)(x - 3)}{x(x - 9)} \times \frac{x^6}{(x + 3)(x + 3)} \div \frac{(2x + 1)(x + 3)}{(2x - 1)(2x + 1)}$$

$$x \neq 0, 9, -3, 3, \tfrac{1}{2}, -\tfrac{1}{2}$$

$$= \frac{\cancel{(x + 3)}(x - 3)}{\cancel{x}(x - 9)} \times \frac{\overset{x^5}{\cancel{x^6}}}{\cancel{(x + 3)}(x + 3)} \times \frac{(2x - 1)\cancel{(2x + 1)}}{\cancel{(2x + 1)}(x + 3)}$$

Never remultiply out the top or bottom unless your life depends on it.

$$= \frac{x^5(x - 3)(2x - 1)}{(x - 9)(x + 3)^2}$$

ADDING AND SUBTRACTING

To be perfectly honest, many students in the past have had difficulty with this section. Please read and reread all these examples very closely. It probably is a good idea to write out all the steps. Be sure you understand each and every step. Then try the exercises. Good luck!!! I'm sure if you practice enough, you will be fine!!

You might want to look at Appendix 1 before or while doing this section. Adding is the same as subtracting: if the bottoms are the same, add (or subtract) the tops. In letters,

$$\frac{a}{d} \pm \frac{b}{d} = \frac{a \pm b}{d}$$

EXAMPLE 1—

$$\frac{4x + 5y}{7} + \frac{2x - 7y}{7}$$

ANSWER

$$\frac{6x - 2y}{7}$$

EXAMPLE 2—

$$\frac{3x + 11}{6} - \frac{3 - x}{6}$$

$$\frac{(3x + 11)}{6} - \frac{+(3 - x)}{6} = \frac{(3x + 11)}{6} + \frac{-(3 - x)}{6}$$

If you need fewer steps, good!!

$$= \frac{(3x + 11) - (3 - x)}{6} = \frac{3x + 11 - 3 + x}{6} = \frac{4x - 8}{6}$$

$$= \frac{4(x - 2)}{6} = \frac{2(x - 2)}{3}$$

If the bottoms are different, you must proceed as you would with large-number bottoms.

1. Factor all denominators.

2. The least common denominator (LCD) is the product of the most number of times a prime factor occurs in any *one* denominator.

3. Multiply top and bottom by what's missing.

4. Simplify and combine all tops.

5. Factor the top and reduce if possible.

EXAMPLE 3

$$\frac{2a - 3b}{6a} - \frac{4a - 2b}{9b}$$

$$6a = 2(3)a \qquad 9b = 3(3)(b)$$

LCD: most number of 2s in any one bottom, 1; most number of 3s, 2; 1 a and 1 b. LCD = (2)(3)(3)ab.

$$\frac{(2a - 3b)}{(2)(3)a} - \frac{+(4a - 2b)}{(3)(3)b} = \frac{3b(2a - 3b)}{(3b)(2)(3)(a)} + \frac{-2a(4a + 2b)}{2a(3)(3)(b)}$$

$$\qquad\qquad\qquad\qquad\qquad 3(b) \text{ missing} \qquad 2(a) \text{ missing}$$

$$= \frac{6ab - 9b^2 - 8a^2 - 4ab}{18ab} = \frac{-8a^2 + 2ab - 9b^2}{18ab}$$

EXAMPLE 4

$$\frac{7}{6ab^3} + \frac{5}{8a^2b}$$

$$6ab^3 = (2)3abbb \qquad 8a^2b = (2)(2)2aab$$

LCD has 3 2s, 1 3, 2 a's, and 3 b's.
LCD = (2)(2)(2)(3)aabbb.

$$\frac{7}{(2)3abbb} + \frac{5}{(2)(2)2aab} = \frac{7(2)(2)a}{(2)(2a)2(3)abbb}$$

$$\qquad\qquad\qquad\qquad\qquad (2)2a \text{ missing}$$

$$+ \frac{5(3bb)}{(3bb)(2)(2)2aab} = \frac{28a + 15b^2}{24a^2b^3}$$

$$\quad 3bb \text{ missing}$$

Okay, let's try some more. Please practice these enough. They are very important.

EXAMPLE 5—

$$\frac{2x}{x^2 - 4} - \frac{3}{x^2 + 4x + 4}$$

$$\frac{2x}{x^2 - 4} + \frac{-3}{x^2 + 4x + 4} = \frac{2x}{(x + 2)(x - 2)} + \frac{-3}{(x + 2)(x + 2)}$$

LCD is most number of times a prime appears in any *one* denominator.

The most number of times $(x - 2)$ in any one denominator is 1.

The most number of times $(x + 2)$ in any ONE denominator is 2. LCD = $(x - 2)(x + 2)(x + 2)$!

In problems like this, if you don't find the LCD, the problems are all much longer. Some you may never be able to fully simplify. We now multiply top and bottom by what's missing.

$$\frac{2x(x + 2)}{(x - 2)(x + 2)(x + 2)} + \frac{-3(x - 2)}{(x - 2)(x + 2)(x + 2)}$$

$$\qquad x + 2 \text{ missing} \qquad\qquad x - 2 \text{ missing}$$

$$= \frac{2x^2 + 4x - 3x + 6}{(x - 2)(x + 2)^2} = \frac{2x^2 + x + 6}{(x - 2)(x + 2)^2}$$

Because the top cannot factor, we are finished.

Let's try a long one.

EXAMPLE 6—

$$\frac{x}{x^2 - 3x - 4} - \frac{1}{x^2 + x} - \frac{4}{x^2 - 4x}$$

$$\frac{x}{(x + 1)(x - 4)} + \frac{-1}{x(x + 1)} + \frac{-4}{x(x - 4)}$$

LCD $= x(x + 1)(x - 4)$

$$= \frac{(x)(x)}{x(x + 1)(x - 4)} + \frac{-1(x - 4)}{x(x + 1)(x - 4)} + \frac{-4(x + 1)}{x(x + 1)(x - 4)}$$

$$= \frac{x^2 - x + 4 - 4x - 4}{x(x + 1)(x - 4)}$$

$$= \frac{x^2 - 5x}{x(x + 1)(x - 4)} = \frac{x(x - 5)}{x(x + 1)(x - 4)} = \frac{(x - 5)}{(x + 1)(x - 4)}$$

Restrictions: $x \neq 0, -1, 4$

Notice that the word *long* was used, not *hard. Long* and *hard* are not the same. There are short easy problems and short hard problems; long easy problems and long hard problems. For completeness, let us do one more.

EXAMPLE 7—

$$3x + 1 + \frac{3}{(x + 2)} \qquad \text{LCD} = (x + 2)$$

$$\frac{(3x + 1)}{1} + \frac{3}{(x + 2)} = \frac{(3x + 1)(x + 2)}{1(x + 2)} + \frac{3}{(x + 2)}$$

$$= \frac{3x^2 + 7x + 2 + 3}{(x + 2)} = \frac{3x^2 + 7x + 5}{x + 2} \qquad x \neq -2$$

COMPLEX FRACTIONS

Sometimes we have fractions on the top and bottom of a fraction. Such a fraction is called *complex.* We would like to simplify them. The easier (and shorter) way to do them is to multiply every fraction top and bottom by the LCD of all the fractions, simplify them, and reduce if possible. Doesn't sound too bad, and it isn't.

EXAMPLE I—

Simplify

$$\dfrac{\dfrac{7}{6}+\dfrac{3}{8}}{4-\dfrac{3}{4}} \qquad \text{LCD}=24$$

This is the answer because it cannot be reduced.

$$\dfrac{\dfrac{24}{1}\cdot\dfrac{7}{6}+\dfrac{24}{1}\cdot\dfrac{3}{8}}{\dfrac{24}{1}\cdot\dfrac{4}{1}-\dfrac{24}{1}\cdot\dfrac{3}{4}}=\dfrac{28+9}{96-18}=\dfrac{37}{78}$$

Hmmm. Not too bad. The algebraic ones aren't too bad either.

EXAMPLE 2—

Simplify

$$\dfrac{\dfrac{1}{x}-\dfrac{1}{y}}{\dfrac{1}{x^2}-\dfrac{1}{y^2}} \qquad \text{LCD}=x^2y^2$$

$$\dfrac{\dfrac{x^2y^2}{1}\cdot\dfrac{1}{x}-\dfrac{x^2y^2}{1}\cdot\dfrac{1}{y}}{\dfrac{x^2y^2}{1}\cdot\dfrac{1}{x^2}-\dfrac{x^2y^2}{1}\cdot\dfrac{1}{y^2}}=\dfrac{xy^2-x^2y}{y^2-x^2}$$

$$=\dfrac{xy(y-x)}{(y-x)(y+x)}=\dfrac{xy}{x+y}$$

where $x \neq 0$

$y \neq 0$

$x \neq y$

$x \neq -y$

EXAMPLE 3

Simplify

$$\frac{\dfrac{2x}{x^2 - 9}}{\dfrac{4}{x + 3} + \dfrac{4}{x - 3}}$$

Because $x^2 - 9 = (x - 3)(x + 3)$, the LCD is $(x - 3)(x + 3)$.

$$\frac{\dfrac{2x}{x^2 - 9}}{\dfrac{4}{x + 3} + \dfrac{4}{x - 3}}$$

$$= \frac{\dfrac{(x - 3)(x + 3)}{1} \cdot \dfrac{2x}{(x - 3)(x + 3)}}{\dfrac{(x - 3)(x + 3)}{1} \cdot \dfrac{4}{(x + 3)} + \dfrac{(x - 3)(x + 3)}{1} \cdot \dfrac{4}{(x - 3)}}$$

$$= \frac{2x}{4(x - 3) + 4(x + 3)} = \frac{2x}{8x} = \frac{1}{4} \qquad x \neq 3, -3, 0$$

This section would not be complete without mentioning a messy kind of complex fraction in which you repeatedly add (or subtract) and simplify. It does not occur too often but it does happen. So let's do one of these. (It is better done a different way.)

EXAMPLE 4

Simplify

$$4 + \frac{3}{2 + \dfrac{1}{x + 3}}$$

$$4 + \cfrac{3}{\cfrac{2}{1} + \cfrac{1}{(x+3)}} = 4 + \cfrac{3}{\cfrac{2}{1} \cdot \cfrac{(x+3)}{(x+3)} + \cfrac{1}{(x+3)}}$$

$$= 4 + \cfrac{\cfrac{3}{1}}{\cfrac{(2x+7)}{(x+3)}}$$

$$= 4 + \cfrac{\cfrac{(x+3)}{1} \cdot \cfrac{3}{1}}{\cfrac{(x+3)}{1} \cdot \cfrac{(2x+7)}{(x+3)}} = \frac{4}{1} + \frac{3x+9}{2x+7}$$

$$= \frac{4(2x+7)}{(2x+7)} + \frac{(3x+9)}{(2x+7)} = \frac{11x+37}{2x+7}$$

Whew!!!

LONG DIVISION

There is no really good place for this, soooo let's do it here. Long division means dividing by a binomial or longer polynomial.

Several things are interesting about long division. First, it's long. But somehow most students like it and get it correct. In fact, although most of you think division is the worst when it comes to arithmetic, with algebra, division is often the one that students would choose. Next, algebraic division is the same as arithmetic division. So let's do a problem very sloooowly.

Divide 6 into 13.

Multiply (2)(6). Subtract 13 − 12 = 1.

Bring down the 5. Divide 6 into 15.

Multiply (6)(2). Subtract 15 − 12 = 3.

Bring down the 7. Divide 6 into 37.

Multiply (6)(6). Subtract 37 − 36 = 1.

Add the remainder, 1, over the divisor, 6. (We write it like this for the algebraic problems.)

$$
\begin{array}{r}
2\ 2\ 6 + 1/6 \\
6\overline{)1\ 3\ 5\ 7} \\
-1\ 2 \\
\hline
1\ 5 \\
-1\ 2 \\
\hline
3\ 7 \\
-3\ 6 \\
\hline
1
\end{array}
$$

Divide, multiply, subtract, bring down. Divide, multiply, subtract, bring down. . . . Almost like a poem. Let's do the algebra.

EXAMPLE 1—

Divide highest power into highest power, 12x³/2x.

Multiply 6x²(2x − 4).

Subtract (12x³ − 30x²) − (12x² − 24x²).

Bring down −4x. Divide 2x into −6x².

Multiply −3x(2x − 4).

Subtract (−6x² − 4x) − (−6x² + 12x).

Bring down +35. Divide 2x into −16x.

Multiply −8(2x − 4).

Subtract (−16x + 35) − (−16 + 32).

The remainder is +3 over the divisor (2x − 4).

$$
\begin{array}{r}
6x^2 - \ 3x \ - 8 \ + 3/(2x-4) \\
2x - 4\overline{)12x^3 - 30x^2 - 4x + 35} \\
\underset{\oplus}{\overline{} } 12x^3 \underset{\ominus}{\overset{+}{}} 24x^2 \\
\hline
-6x^2 - 4x \\
\underset{\ominus}{\overset{+}{}}6x^2 \underset{\oplus}{\overset{-}{}} 12x \\
\hline
-16x + 35 \\
\underset{\ominus}{\overset{+}{}}16x \underset{\oplus}{\overset{-}{}} 32 \\
\hline
+3
\end{array}
$$

This is really not too bad. It looks much worse than it is. Try it. . . . Let's do a more complicated one.

EXAMPLE 2—

Divide $5x^3 - 6x + 2x^5 + 7 + 4x^4$ by $2 + x^3 + x^2$.

1. Arrange the divisor highest power to lowest power: $x^3 + x^2 + 2$.

2. Arrange the dividend (that's what it's called), highest power to lowest. Fill in any missing power by 0 times that term (you'll see why!!): $2x^5 + 4x^4 + 5x^3 + 0x^2 - 6x + 7$.

3. Divide highest power into highest power (same as before).

4. Multiply. Be sure to line up like terms.

5. Subtract and bring down the next term of the dividend.

6. Repeat steps.

7. The problem ends when the remainder is a lower degree than the divisor.

$$2x^2 + 2x + 3 + \frac{-7x^2 - 10x + 1}{x^3 + x^2 + 2}$$

Divide x^3 into $2x^5$
Multiply $2x^2 (x^3 + x^2 + 2)$.
Line up like terms. That's
why we put $0x^2$. Subtract
$(2x^5 + 4x^4 + 5x^3 + 0x^2) -$
$(2x^5 + 2x^4 + 4x^2)$.
Bring down $-6x$. Divide x^3
into $2x^4$.
Multiply $2x(x^3 + x^2 + 2)$.
Subtract $(2x^4 + 5x^3 -$
$4x^2 - 6x) - (2x^4 + 2x^3 - 4x)$.
Bring down the 7. Divide
x^3 into $3x^3$.
Multiply $3(x^3 + x^2 + 2)$.
Subtract $(3x^3 - 4x^2 -$
$2x + 7) - (3x^3 + 3x^2 + 6)$.
Problem is over because
the $-7x^2 - 10x + 1$ is
lower degree than $x^3 +$
$x^2 + 2$.

$$x^3 + x^2 + 2 \overline{\smash{)}2x^5 + 4x^4 + 5x^3 + 0x^2 - 6x + 7}$$
$$\underline{\overset{\ominus}{}\,2x^5 \overset{\ominus}{}\,2x^4 \qquad\quad \overset{\ominus}{}\,4x^2}$$

$$2x^4 + 5x^3 - 4x^2\ - 6x$$

$$\underline{\overset{\ominus}{}\,2x^4 \overset{\ominus}{}\,2x^3 \qquad\quad \overset{\ominus}{}\,4x}$$

$$3x^3 - 4x^2 - 10x + 7$$

$$\underline{\overset{\ominus}{}\,3x^3 \overset{\ominus}{}\,3x^2 \qquad\quad \overset{\ominus}{}\,6}$$

$$-7x^2 - 10x + 1$$

When you get good at this, you won't have to put in terms like $0x^2$.

FRACTIONAL EQUATIONS

We've done some of these before. Let's do more now.

EXAMPLE 1—

Solve for x:

I know; we did some of
these before. If you know
the first ones, skip them!

$$\frac{x-6}{2} + \frac{5x+5}{3} = \frac{x}{4} + 13 \qquad \text{LCD} = 12$$

$$\frac{12(x-6)}{2} + \frac{12(5x+5)}{3} = \frac{12x}{4} + 12(13)$$

$$6x - 36 + 20x + 20 = 3x + 156$$

$$26x - 16 = 3x + 156$$

$23x = 172$

$$x = \frac{172}{23}$$

You need not check this unless you want or your teacher wants.

EXAMPLE 2

Solve for x:

$$\frac{2}{3x} + \frac{4}{5x} = \frac{11}{15} \qquad \text{LCD} = 15x$$

$$\frac{15x}{1}\left(\frac{2}{3x}\right) + \frac{15x}{1}\left(\frac{4}{5x}\right) = \frac{15x}{1}\left(\frac{11}{15}\right)$$

$10 + 12 = 11x$

$11x = 22$

$x = 2$

You *must* check if there is a variable in the denominator, to make sure the denominator does not equal 0.

$$\frac{2}{3(2)} + \frac{4}{5(2)} \overset{?}{=} \frac{11}{15}$$

$$\frac{1}{3} + \frac{2}{5} = \frac{11}{15}$$

And the problem checks.

EXAMPLE 3

Solve for y:

$$\frac{5}{y-8} = \frac{8}{y-5}$$

$8(y-8) = 5(y-5)$ **Cross multiply.**

$8y - 64 = 5y - 25$

$3y = 39$

$y = 13$

Unless we made a mistake, 13 must be an answer, because the only bad answers are 5 and 8. (Why?)

EXAMPLE 4—

Solve for z:

$$\frac{4}{z - 2} = \frac{z}{6}$$

$$z(z - 2) = 24$$

$$z^2 - 2z - 24 = 0$$

$$(z - 6)(z + 4) = 0$$

Soooo, $z = 6$ and $z = -4$. (Check both; really easy.)

EXAMPLE 5—

Solve for x:

$$\frac{6}{x - 1} = \frac{12}{x^2 - 1} - 2$$

$x^2 - 1 = (x + 1)(x + 1)$, sooo LCD = $(x + 1)(x - 1)$.

$$(x + 1)(x - 1)\,\frac{6}{x - 1} = (x + 1)(x - 1)\,\frac{12}{(x + 1)(x - 1)}$$

$$- (x + 1)(x - 1)(2)$$

$$6(x + 1) = 12 - 2(x + 1)(x - 1)$$

$$6x + 6 = 12 - 2x^2 + 2$$

$$2x^2 + 6x - 8 = 0$$

Sooooo

$$2(x^2 + 3x - 4) = 2(x + 4)(x - 1) = 0$$

$$x = -4 \qquad \text{or} \qquad x = 1$$

Let us check.

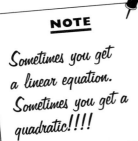

NOTE

Sometimes you get a linear equation. Sometimes you get a quadratic!!!!

$$\frac{6}{-4-1} \overset{?}{=} \frac{12}{(-4)^2 - 1} - 2$$

$$\frac{-6}{5} \overset{?}{=} \frac{4}{5} - 2$$

$$\frac{-6}{5} = \frac{4}{5} - \frac{10}{5}$$

And the problem checks. Buuuuuuuut,

$$\frac{6}{1-1} \overset{?}{=} \frac{12}{1^2 - 1} - 2$$

$$\frac{6}{0} \overset{?}{=} \frac{12}{0} - 2$$

Since you cannot divide by 0, 1 is not a root. The only root is x = −4. Sometimes both will check; sometimes one will check; sometimes none will check, in which case there is no solution.

WORD PROBLEMS WITH FRACTIONS

Age Problems Revisited

EXAMPLE 1—

Sandy is four times Brett's age. Six years ago, Brett was 1/7 Sandy's age. Find their ages now.

We will do it by chart. Let x = Brett's age (the younger).
4x is Sandy's age. The second column is the age 6
years ago (−6).

	Age now	Age 6 years ago
Brett	x	x − 6
Sandy	4x	4x − 6

6 years ago, Brett (x − 6) was (=) 1/7 Sandy's age
[(1/7)(4x − 6)].

$$\frac{x-6}{1} = \frac{1}{7}(4x - 6)$$

$$7(x - 6) = 4x - 6$$

$$7x - 42 = 4x - 6$$

$$3x = 36$$

$$x = 12 \qquad \text{Brett's age now}$$

$$4x = 48 \qquad \text{Sandy's age now}$$

EXAMPLE 2—

Cory is twice as old as Bobby. In 12 years Bobby will
be 7/12 as old as Cory. Find their ages now.

	Age now	Age in 12 years
Bobby	x	x + 12
Cory	2x	2x + 12

$$x + 12 = \frac{7}{12}(2x + 12)$$

$$12(x + 12) = 7(2x + 12)$$

$$12x + 144 = 14x + 84$$

$$-2x = -60$$

$$x = 30 \qquad \text{Bobby's age now}$$

$$2x = 60 \qquad \text{Cory's age now}$$

Work Problems

THEORY

A job can be done in 20 hours. In 1 hour, 1/20 is done. In 7 hours, 7/20 is done.

EXAMPLE 1

Lee can do a job in 6 hours and Pat can do the same job in 12 hours. Working together, how many hours to do the whole job?

I always tell my students there are two answers. Lee brings over a case of soda, chips, pretzels, a radio, and 10 friends, and the job never gets done. But seriously, let's do the problem. Let x = the number of hours to do the job together. x/6 is the part done by Lee and x/12 is the part done by Pat. The part done by Pat plus the part done by Lee equals the whole job!!!!

$$\frac{x}{6} + \frac{x}{12} = 1 \qquad \text{(One whole job!)}$$

$$2x + x = 12$$

$$x = 4 \text{ hours}$$

EXAMPLE 2

Cameron does a job in 18 hours. At the start Blair joins, and they finish the job in 12 hours. How many hours does it take Blair to do the job alone?

Let x = number of hours for Blair to do the job. In 1 hour Blair does 1/x job. In 12 hours, Blair does 12/x job. In 12 hours Cameron does 12/18 job.

$$12/x + 12/18 = 1$$

Solving, we get x = 36.

Distance Problems Revisited

EXAMPLE 1

A car goes 600 miles in the same time that a car going 25 mph less goes 400 miles. Find each of their speeds.

If rate times time equals distance, then time = distance/rate. "In the same time as" means the times are equal.

	Rate	**Time = d/r**	**Distance**
Slower	x − 25	400/(x − 25)	400
Faster	x	600/x	600

$$\frac{600}{x} = \frac{400}{x - 25}$$

$600x - 15{,}000 = 400x$

$200x = 15{,}000$

$x = 75$ mph faster

$x - 25 = 50$ mph slower

EXAMPLE 2

A train makes the 800-mile trip between two cities. On the return trip, it takes 2 hours more because the speed is 20 mph less. Find the speed of each trip.

$r \times t = d$, soooo $r = d/t$. t = hours of first trip; $t + 2$ = hours of the return trip.

	r	**×**	**t**	**=**	**d**
Trip (faster)	800/t		t		800
Return (slower)	800/(t + 2)		t + 2		800

$$\frac{800}{t + 2} + 20 = \frac{800}{t} \qquad \text{LCD} = t(t + 2)$$

Slower speed + 20 = faster speed

$800t + 20t(t + 2) = 800(t + 2)$

$800t + 20t^2 + 40t = 800t + 1600$

$20t^2 + 40t - 1600 = 0$

$20(t^2 + 2t - 80) = 20(t + 10)(t - 8) = 0$

$t = -10$? Wrong answer—there is no such thing as negative time!!! $t = 8$ hours is the answer.

Sooo . . . .

Trip (faster): $r = \dfrac{d}{t} = \dfrac{800 \text{ mi}}{8 \text{ hrs}} = 100$ mph

Return (slower): $r = \dfrac{d}{t} = \dfrac{800 \text{ mi}}{8 + 2 \text{ hrs}} = 80$ mph

CHAPTER 7

RADICALS AND EXPONENTS

When I started teaching, I thought that this chapter, especially the beginning, caused students many problems. Maybe it was me, and maybe I was wrong, but now my students learn this chapter, especially radicals, super well. I'm sure you will too.

DEFINITION

$\sqrt{a}$, "square root of a": the number b, such that (b)(b) = a.

$\sqrt{9} = 3$, because (3)(3) = 9. $\sqrt{25} = 5$ because (5)(5) = 25. 9 and 25 are called *perfect squares* because they have exact square roots.

$\sqrt[3]{a}$, "cube root of a": the number c, such that (c)(c)(c) = a. $\sqrt[3]{64} = 4$ because (4)(4)(4) = 64.

$\sqrt[4]{a}$, "the fourth root of a": the number h, such that (h)(h)(h)(h) = a. $\sqrt[4]{81} = 3$ because (3)(3)(3)(3) = 81.

$\sqrt[n]{a}$, "the nth root of a": the number b such that (b)(b)(b) . . . (b) = a (n factors).

EXAMPLE 1—

$\sqrt{16} = 4$; $-\sqrt{16} = -4$ (minus sign is on the outside of square root).

$\pm\sqrt{16} = \pm 4$ ($\pm\sqrt{16}$ means two numbers, $+\sqrt{16}$ and $-\sqrt{16}$).

$\sqrt{-16}$ is not a real number because a positive times a positive is positive and a negative times a negative is positive. No real number squared gives a negative. Also, $\sqrt{0} = 0$. Buuuut . . .

EXAMPLE 2—

$\sqrt[3]{1000} = 10$; $\sqrt[3]{-125} = -5$; $\sqrt[3]{0} = 0$

$-\sqrt[3]{-27} = -(-3) = 3$; $\pm\sqrt[3]{216} = \pm 6$; $\pm\sqrt[3]{-8} = \mp 2$

(The chart is on the next page.)

As you do these problems, you should keep this list in front of you. Don't go home and memorize. But as you do the problems, you will be surprised how many you learn from repetition.

EXAMPLE 3—

Solve for x:

$x^2 - 16 = 0$

$x^2 = 16$

Two roots.

$x = \pm\sqrt{16} = \pm 4$

We will do more of these in another chapter.

Powers

	Square	Cube	4th	5th	6th	7th	8th	9th	10th	11th	12th
2	4	8	16	32	64	128	256	512	1024	2048	4096
3	9	27	81	243	729						
4	16	64	256	1024							
5	25	125	625	3125							
6	36	216									
7	49	343									
8	64	512									
9	81	729									
10	100	1000									
11	121	1331									
12	144	1728									
13	169										
14	196										
15	225										
16	256										
17	289										
18	324										
19	361										
20	400										
21	441										
22	484										
23	529										
24	576										
25	625										
26	676										
27	729										
28	784										
29	841										
30	900										
31	961										
32	1024										

The chart is read backwards and forwards. 5 cubed is
125, buuuut also the cube *root* of 125 is 5.

You will find that most of these numbers repeat over and over; some now and some later on.

Try not to use calculators. If you put these numbers in your brain, many things go easier for you.

Some numbers are not perfect squares. 2 is not a perfect square. There is no integer or rational number that gives you 2 when you square it. $\sqrt{2}$ is *irrational*.

An *irrational number* is any number that cannot be written as a terminating or repeating decimal (any number that is not rational).

π, $\sqrt{7}$, and $\sqrt[3]{5}$ are all examples of irrational numbers.

Remember, writing 3.14 or 22/7 are only approximations for π.

A *real* number is any decimal number.

NOTES

We have only talked about real numbers to this point.

The real numbers are the rationals plus the irrationals.

We will talk more rationals, irrationals, and reals soon.

Are there numbers that are not real? Yes!! We will learn about them in this chapter, and they are very easy, as you will see.

Now that we know what square and other roots are, we would like to simplify them; add, subtract, multiply, and divide them; and apply them.

95 percent of the time, the shortest way is the simplest. I was always taught to use the KISS method of teaching— keep it simple, stupid. However, the radical unit is one

of the few times we don't use the shortest method. The method is short, just not the shortest.

SIMPLIFYING RADICALS

EXAMPLE 1—

Simplify

$\sqrt{18}$

We write 18 as the product of primes: 18 = 2(3)(3).

$= \sqrt{2\,\boxed{(3)(3)}}$

$= 3\sqrt{2}$

Two 3s circled on the inside become one 3 on the outside. Why? Because $\sqrt{(3)(3)} = (\sqrt{9}) = 3$.

EXAMPLE 2—

Simplify

$\sqrt{72}$

$= \sqrt{\boxed{(2)(2)}\,(2)\,\boxed{(3)(3)}} = 2 \times 3\sqrt{2} = 6\sqrt{2}$

Break 72 into primes. No matter how you do it, you will get the same ones.

EXAMPLE 3—

Simplify

$10\sqrt{125}$

$= 10\sqrt{\boxed{(5)(5)}\,(5)} = 10(5)\sqrt{5} = 50\sqrt{5}$

EXAMPLE 4—

Simplify

$\sqrt{27b^7}$

$= \sqrt{\boxed{(3)(3)}\,(3)\,\boxed{(b)(b)}\,\boxed{(b)(b)}\,\boxed{(b)(b)}\,(b)} = 3b^3\sqrt{3b}$

EXAMPLE 5—

Simplify

$\sqrt{a^{11}}$

$= \sqrt{\overline{aa}\,\overline{aa}\,\overline{aa}\,\overline{aa}\,\overline{aa}\,a} = a^5\sqrt{a}$

Short way!!!! Nice! $\sqrt{}$ means $\sqrt[2]{}$. Divide 2 into 11. We get 5 groups of 2. So a^5 goes on the outside. The remainder is 1. a goes on the inside. That's all!

EXAMPLE 6—

Simplify

$4y^3\sqrt{25y^3}$

$= (4y^3)(5y)\sqrt{y} = 20y^4\sqrt{y}$

EXAMPLE 7—

Simplify

$\sqrt[3]{72a^{14}b^9}$

With cube roots, we circle 3 of the same factors. (Fourth roots we circle 4 of the same.) 3 into 14 is 4 with a remainder of 2. 3 into 9 is 3 with no remainder. Soooooo,

$\sqrt[3]{72a^{14}b^9} = 2a^4b^3\sqrt[3]{(3)(3)a^2} = 2a^4b^3\sqrt[3]{9a^2}$

Since $\sqrt[3]{72} = \sqrt[3]{\overline{(2)\,(2)\,(2)}\,(3)(3)} = 2\sqrt[3]{(3)(3)}$

Not too bad!!

ADDING AND SUBTRACTING RADICALS

Adding and subtracting radicals is the same as like terms in Chap. 1. With like radicals, you add or subtract the coefficients. Unlike radicals cannot be combined.

EXAMPLE 1—

Simplify

$4\sqrt{2} - 5\sqrt{7} - 7\sqrt{2} - 10\sqrt{7}$

The answer issss

$-3\sqrt{2} - 15\sqrt{7}$

EXAMPLE 2—

Simplify

$3\sqrt{27} + 5\sqrt{8} + 6\sqrt{12}$

We must simplify each radical first.

$= 3\sqrt{(3)(3)\,(3)} + 5\sqrt{(2)(2)\,(2)} + 6\sqrt{(2)(2)\,(3)}$

$= (3)(3)\sqrt{3} + 5(2)\sqrt{2} + (6)(2)\sqrt{3} = 9\sqrt{3} + 10\sqrt{2} + 12\sqrt{3}$

$= 21\sqrt{3} + 10\sqrt{2}$

EXAMPLE 3—

Simplify

$\sqrt[3]{128y^8} + 10y\sqrt[3]{54y^5}$

$= \sqrt[3]{(2)(2)(2)\,(2)(2)(2)\,(2)\,(y)(y)(y)(y)(y)(y)\,(y)(y)}$

$+ 10y\sqrt[3]{(2)\,(3)(3)(3)\,(y)(y)(y)\,(y)(y)}$

$= 4y^2\sqrt[3]{2y^2} + 10y(3y)\sqrt[3]{2y^2} = 34y^2\sqrt[3]{2y^2}$

MULTIPLYING RADICALS

There are 2 types. Sometimes they occur together.

If $a, b \geq 0$, then $\sqrt{a} \times \sqrt{b} = \sqrt{ab}$. In fact, if n is positive even, $\sqrt[n]{a} \times \sqrt[n]{b} = \sqrt[n]{ab}$. If n is odd positive, then a and b can be any real number and $\sqrt[n]{a} \times \sqrt[n]{b} = \sqrt[n]{ab}$. In particular, $\sqrt[3]{a} \times \sqrt[3]{b} = \sqrt[3]{ab}$.

Also, $a\sqrt{b} \times c\sqrt{d} = ac\sqrt{bd}$. You multiply the insides and multiply the outsides. Again, both b and d must be greater than or equal to 0.

EXAMPLE 1—

Multiply

$3\sqrt{6} \times 5\sqrt{7}$

The answer is $15\sqrt{42}$.

EXAMPLE 2—

Multiply (the easier way)

$7\sqrt{8} \times 10\sqrt{6}$

$= 70\sqrt{(2)(2)(2)(2)(3)}$

$= (70)(2)(2)\sqrt{3} = 280\sqrt{3}$.

If you first multiplied 6 times 8, the first thing you would have to do is break it down again. So write each one in terms of primes first to avoid wasting effort.

EXAMPLE 3—

$4\sqrt{2}(3\sqrt{5} + \sqrt{7} - 5 + 8\sqrt{6} + 3\sqrt{2})$

$= 12\sqrt{10} + 4\sqrt{14} - 20\sqrt{2} + 32\sqrt{12} + 12\sqrt{4}$

Multiply outsides Multiply insides	Multiply insides only	Multiply outsides only	Multiply outsides Multiply insides	Multiply outsides Multiply insides
			$\sqrt{12} = 2\sqrt{3}$	$\sqrt{4} = 2$

$= 12\sqrt{10} + 4\sqrt{14} - 20\sqrt{2} + 64\sqrt{3} + 24$

If you couldn't combine inside the parentheses before multiplication, you won't be able to combine terms after the multiplication.

EXAMPLE 4—

Multiply

$(2\sqrt{3} + 4\sqrt{2})(6\sqrt{3} + 2\sqrt{2})$

This looks familiar. Aha!! We must FOIL:

$(2\sqrt{3} + 4\sqrt{2})(6\sqrt{3} + 2\sqrt{2})$

$= 12\sqrt{9} + 4\sqrt{6} + 24\sqrt{6} + 8\sqrt{4}$

$= 36 + 28\sqrt{6} + 16 = 52 + 28\sqrt{6}$

Or, to be fancy,

$4(13 + 7\sqrt{6})$

DIVIDING RADICALS

The last operation is division, but in order to fully do this, we have to rationalize the denominator. Soooo, division will be in the middle of two sections on rationalizing.

Rationalize the Denominator I

Suppose we have $1/\sqrt{2}$. In the past the reason we never wanted radicals in the bottom was the problem of finding an approximate decimal value for $1/\sqrt{2}$. $\sqrt{2}$ is approximately 1.414 (which you should know, as you should $\sqrt{3}$, which is the year George Washington was born—1.732). To find the value of $1/\sqrt{2} = 1/1.414$, we had to divide $1.414\overline{)1.000000}$. This had two great disadvantages: this is very long division, and it is inaccurate because the divisor, 1.414, is rounded off. Now suppose we rationalize the denominator. Look what happens. . . .

EXAMPLE I—

Rationalize

$$\frac{1}{\sqrt{2}} = \frac{1}{\sqrt{2}} \times \frac{\sqrt{2}}{\sqrt{2}} = \frac{\sqrt{2}}{2}$$

By multiplying the top and bottom by $\sqrt{2}$, the arithmetic is much easier to do because . . . if you take $\sqrt{2}/2 = 1.414/2 = 2\overline{)1.414}$, first you are dividing by 2 (very easy) and second, when you round off the top, it is more accurate. However, we really don't have to do this with calculators. In math we almost always have to rationalize the numerator. *Math is behind the times.* I have found that when math is behind the times, math teaching is good. Unfortunately, as I write this book, math is the most progressive subject. In my view, math teaching today is at its worst and so is math learning. This is why I wrote this book!!!! Enough talk. I talk too much. Let's go on!!

EXAMPLE 2—

Rationalize the denominator

$$\frac{7}{\sqrt{75}}$$

Simplify the bottom first!!!

$$\sqrt{75} = \sqrt{3(5)(5)} = 5\sqrt{3}$$

Soooo,

$$\frac{7}{\sqrt{75}} = \frac{7}{5\sqrt{3}} = \frac{7}{5\sqrt{3}} \times \frac{\sqrt{3}}{\sqrt{3}} = \frac{7\sqrt{3}}{15}$$

EXAMPLE 3—

Rationalize the denominator

$$\frac{9}{\sqrt{8a^5}} = \frac{9}{2a^2\sqrt{2a}} = \frac{9}{2a^2\sqrt{2a}} \frac{\sqrt{2a}}{\sqrt{2a}} = \frac{9\sqrt{2a}}{4a^3}$$

Not too bad. Let's do division!!!

Division

Division is based on one simple principle:

$$\sqrt{\frac{a}{b}} = \frac{\sqrt{a}}{\sqrt{b}}$$

where a is not negative, buuuut b must be positive!

Let's do a bunch of problems!!!

EXAMPLE 1

$$\sqrt{\frac{4}{9}} = \frac{\sqrt{4}}{\sqrt{9}} = \frac{2}{3}$$

Really easy.

EXAMPLE 2

$$\sqrt{\frac{3}{25}} = \frac{\sqrt{3}}{\sqrt{25}} = \frac{\sqrt{3}}{5}$$

EXAMPLE 3

$$\sqrt{\frac{7}{27}} = \frac{\sqrt{7}}{\sqrt{27}} = \frac{\sqrt{7} \times \sqrt{3}}{3\sqrt{3} \times \sqrt{3}} = \frac{\sqrt{21}}{9}$$

Notice that once you divide, the problems are just like those in the first part of this section.

EXAMPLE 4

$$\sqrt{\frac{2a^3b^7}{6a^{10}b^2}} = \sqrt{\frac{1b^5}{3a^7}} = \frac{\sqrt{b^5}}{\sqrt{3a^7}} = \frac{b^2\sqrt{b}}{a^3\sqrt{3a}} \times \frac{\sqrt{3a}}{\sqrt{3a}} = \frac{b^2\sqrt{3ab}}{3a^4}$$

Pretty long, but not bad either. Let's try some more.

EXAMPLE 5

$$\frac{\sqrt{10a^4}}{\sqrt{35a^7}} = \sqrt{\frac{10a^4}{35a^7}} = \sqrt{\frac{2}{7a^3}} = \frac{\sqrt{2}}{\sqrt{7a^3}} = \frac{\sqrt{2} \times \sqrt{7a}}{a\sqrt{7a} \times \sqrt{7a}}$$

$$= \frac{\sqrt{14a}}{7a^2}$$

We make one square root, simplify, and then break it up.

Lastly, let's do one with addition.

EXAMPLE 6—

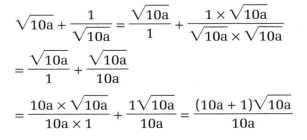

$$\sqrt{10a} + \frac{1}{\sqrt{10a}} = \frac{\sqrt{10a}}{1} + \frac{1 \times \sqrt{10a}}{\sqrt{10a} \times \sqrt{10a}}$$

$$= \frac{\sqrt{10a}}{1} + \frac{\sqrt{10a}}{10a}$$

$$= \frac{10a \times \sqrt{10a}}{10a \times 1} + \frac{1\sqrt{10a}}{10a} = \frac{(10a + 1)\sqrt{10a}}{10a}$$

Rationalize the Denominator II

In the previous parts of this section, when we rationalized the denominator, the bottom was a *monomial* (one term). Let's see what happens when the bottom is a *binomial*.

EXAMPLE I—

Rationalize the denominator

$$\frac{2\sqrt{3}}{4\sqrt{2} + \sqrt{7}}$$

To rationalize, we multiply top and bottom by the *conjugate* of the bottom, $4\sqrt{2} - \sqrt{7}$. To get the conjugate, one changes the sign in the middle.

$$\frac{2\sqrt{3} \times (4\sqrt{2} - \sqrt{7})}{(4\sqrt{2} + \sqrt{7}) \times (4\sqrt{2} - \sqrt{7})} = \frac{8\sqrt{6} - 2\sqrt{21}}{25}$$

Multiply out the top, FOIL the bottom and simplify if possible. The bottom will always be an integer.

EXAMPLE 2—

Rationalize the denominator

$$\frac{\sqrt{c}}{\sqrt{c} - \sqrt{d}}$$

$$\frac{\sqrt{c} \times (\sqrt{c} + \sqrt{d})}{(\sqrt{c} - \sqrt{d}) \times (\sqrt{c} + \sqrt{d})} = \frac{c + \sqrt{cd}}{c - d}$$

Note that both terms don't have to be square roots. The conjugate of $6 + \sqrt{x}$ is $6 - \sqrt{x}$. The conjugate of $\sqrt{y} + 3$ is $\sqrt{y} - 3$.

EXPONENTS

Law of Exponents for Positive Integers

We will finally finish this section. We have done three of the laws and hinted at the other two. Now let's do all five together.

Law 1: $x^m x^n = x^{m+n}$ $\qquad x^6 x^4 = x^{10}$

Law 2: $\dfrac{x^m}{x^n} = x^{m-n}$ $\qquad$ if m is bigger

orrrr

$\dfrac{1}{x^{n-m}}$ $\qquad$ if n is bigger

$\dfrac{y^6}{y^2} = y^4$ $\qquad$ and $\qquad$ $\dfrac{z^4}{z^{100}} = \dfrac{1}{z^{96}}$

Law 3: $(x^m)^n = x^{mn}$ $\qquad (a^6)^4 = a^{24}$
since $(a^6)^4 = a^6 a^6 a^6 a^6$

Law 4: $(ab)^m = a^m b^m$ $\qquad (ab)^3 = a^3 b^3$
since $(ab)^3 = (ab)(ab)(ab)$

Law 5: $\left(\dfrac{a}{b}\right)^n = \dfrac{a^n}{b^n}$ $\qquad b \neq 0$

$\left(\dfrac{a}{b}\right)^4 = \dfrac{a^4}{b^4}$ $\qquad$ since $\left(\dfrac{a}{b}\right)^4 = \left(\dfrac{a}{b}\right) \times \left(\dfrac{a}{b}\right) \times \left(\dfrac{a}{b}\right) \times \left(\dfrac{a}{b}\right)$

We've done some problems with these. Let's do a few more.

EXAMPLE 1—

Simplify

$(4ab^3)^3 = (4a^1b^3)^3 = 4^3a^3b^9 = 64a^3b^9$

Don't forget the exponent 1 on a!!

EXAMPLE 2—

Simplify

$$\left(\frac{4x^4y^7}{6x^9y^3}\right)^2$$

There are two basic ways to do this problem:

METHOD 1

$$\left(\frac{4x^4y^7}{6x^9y^3}\right)^2 = \left(\frac{2y^4}{3x^5}\right)^2 = \frac{4y^8}{9x^{10}}$$

METHOD 2

$$\left(\frac{4x^4y^7}{6x^9y^3}\right)^2 = \left(\frac{16x^8y^{14}}{36x^{18}y^6}\right) = \frac{4y^8}{9x^{10}}$$

Both are correct, but method 1 is probably easier because the numbers stay smaller.

In the next example, we have no choice because the outside exponents are different.

EXAMPLE 3—

Simplify

$$\frac{(4c^2d^3)^3}{(6c^7d)^2} = \frac{64c^6d^9}{36c^{14}d^2} = \frac{16d^7}{9c^8}$$

Law of Exponents for All Integers

The laws for integer exponents are the same as for positive integers. But what the heck is a negative exponent? Negative exponent does not mean we have a negative number. Negative exponent means *reciprocal*.

DEFINITION

$$a^{-n} = \frac{1}{a^n}$$

NOTE

$$\frac{1}{b^{-n}} = b^n \qquad \text{since} \quad \frac{1}{b^{-n}} = \frac{1 \cdot b^n}{\frac{1}{b^n} \cdot \frac{b^n}{1}} = b^n$$

annnd

$$\left(\frac{a}{b}\right)^{-n} = \frac{b^n}{a^n} \qquad \text{since} \quad \left(\frac{a}{b}\right)^{-n} = \frac{a^{-n}}{b^{-n}} = \frac{b^n}{a^n}$$

specifically

$$\left(\frac{c}{d}\right)^{-1} = \frac{d}{c}$$

EXAMPLE 1—

Evaluate the following:

a. 6^{-2} b. 5^{-3} c. $(-2)^{-4}$ d. $(-3)^{-3}$ e. $(4/5)^{-2}$ f. $-4(2)^{-3}$

a. $\frac{1}{6^2} = \frac{1}{36}$ b. $\frac{1}{5^3} = \frac{1}{125}$ c. $\frac{1}{(-2)^4} = \frac{1}{16}$

d. $\frac{1}{(-3)^3} = \frac{1}{-27}$ e. $(5/4)^2 = 25/16$ f. $\frac{-4}{2^3} = \frac{-4}{8} = -\frac{1}{2}$

EXAMPLE 2—

Simplify

$$\frac{a^5 b^{-7} c^9}{a^{-3} b^{-9} c^{40}} = \frac{a^5 a^3 b^9 c^9}{b^7 c^{40}} = \frac{a^8 b^2}{c^{31}}$$

EXAMPLE 3—

Simplify

$$\frac{(10a^4b^{-5})^{-2}}{(a^{-5}b^{-8})^{-3}} = \frac{10^{-2}a^{-8}b^{10}}{a^{15}b^{24}} = \frac{b^{10}}{100a^8a^{15}b^{24}} = \frac{1}{100a^{23}b^{14}}$$

NOTE

If the exponent is negative on top, it is positive on the bottom, and a negative exponent in the bottom is positive on top (positive exponential terms don't change). The reason we can do this is that there is *no* adding or subtracting. The problem becomes longer if we have a + or −.

EXAMPLE 4—

Simplify

A *complex fraction!!!!* LCD is a²b².

$$\frac{a^{-1} + b^{-1}}{a^{-2} - b^{-2}} = \frac{\dfrac{1}{a} + \dfrac{1}{b}}{\dfrac{1}{a^2} - \dfrac{1}{b^2}}$$

$$= \frac{\dfrac{1}{a} \times \dfrac{a^2b^2}{1} + \dfrac{1}{b} \times \dfrac{a^2b^2}{1}}{\dfrac{1}{a^2} \times \dfrac{a^2b^2}{1} - \dfrac{1}{b^2} \times \dfrac{a^2b^2}{1}} = \frac{ab^2 + a^2b}{b^2 - a^2} = \frac{ab(b + a)}{(a - b)(a + b)}$$

$$= \frac{ab}{a - b}$$

Now, let's relate the beginning of the chapter to the middle of the chapter.

Fractional Exponents

We would like to relate fractional exponents to radicals.

DEFINITION

$$x^{1/r} = \sqrt[r]{x}$$

where x is not negative if r is even

$$8^{1/3} = \sqrt[3]{8} = 2 \qquad 25^{1/2} = \sqrt{25} = 5$$

NOTE

$$a^{p/r} = \sqrt[r]{a^p} \text{ or } (\sqrt[r]{a})^p$$

where p = power
          r = root

This says you can do the power first or root first. But which one???? Let's see.

EXAMPLE 1—

Let us do $8^{5/3}$.

$$8^{5/3} = (8^5)^{1/3} = \sqrt[3]{8^5}$$

$(8^5)^{1/3}$ says multiply $(8)(8)(8)(8)(8)$, and then, when you get the answer, take the cube root. I don't think I want to do this. Buuuut

$$8^{5/3} = (\sqrt[3]{8})^5 = 2^5 = 32$$

That's much better. Root before power, all the time!!

EXAMPLE 2—

$$27^{-4/3} = \frac{1}{27^{4/3}} = \frac{1}{(\sqrt[3]{27})^4}$$

$$= \frac{1}{3^4} = \frac{1}{81}$$

Keep writing the one on the top of the fraction or you'll forget.

Now the exponent rules also hold for fractional exponents.

NOTE

Don't forget the number part.

EXAMPLE 3—

Simplify

a. $(x^{3/4})^{5/7} = x^{15/28}$

b. $(4b^6)^{3/2} = 4^{3/2}(b^6)^{3/2} = (\sqrt{4})^3 b^9 = 8b^9$

EXAMPLE 4—

Simplify

$$\frac{x^{4/3}x^{1/4}}{x^{5/6}} = x^{4/3 + 1/4 - 5/6} = x^{16/12 + 3/12 - 10/12} = x^{9/12} = x^{3/4}$$

EXAMPLE 5—

If $x^3 = 64$, find $x^{-1/2}$.

You should know that $4^3 = 64$, $x = 4$. $4^{-1/2} = 1/4^{1/2} = 1/\sqrt{4} = 1/2$.

This is a kind of problem the SAT might ask. Let's try one more.

EXAMPLE 6—

$x^{-3/4} = 1/8$. $x^2 = ?$

$x = (x^{-3/4})^{-4/3} = (1/8)^{-4/3} = (8/1)^{4/3} = (\sqrt[3]{8})^4 = 2^4 = 16$. So $x^2 = 16^2 = 256$.

Let's do a few more examples.

EXAMPLE 7—

Multiply

$$\sqrt[4]{x^3} \times \sqrt[5]{x^6} = x^{3/4} \times x^{6/5} = x^{15/20 + 24/20} = x^{39/20} = \sqrt[20]{x^{39}}$$

EXAMPLE 8—

Rationalize the denominator

$$\frac{1}{\sqrt[5]{a^{11}}}$$

As a fractional exponent this is

$$\frac{1}{a^{11/5}}$$

We have to find the smallest power of "a" (the fraction) that, when you multiply a to that power, gives a positive integer. (When you multiply, you *add* exponents.)

$11/5 + 4/5 = 15/5 = 3$

Soooo,

$$\frac{1 \times a^{4/5}}{a^{11/5} \times a^{4/5}} = \frac{a^{4/5}}{a^{15/5}} = \frac{\sqrt[5]{a^4}}{a^3}$$

EXAMPLE 9—

Rationalize

$$\frac{1}{\sqrt[3]{3}} = \frac{1 \times 3^{2/3}}{3^{1/3} \times 3^{2/3}} = \frac{3^{2/3}}{3} = \frac{(3^2)^{1/3}}{3} = \frac{\sqrt[3]{9}}{3}$$

We will finish this chapter with divisibility, rational, irrational, and real numbers.

NUMBER THEORY

Divisibility

This small section has many different little parts. We begin with divisibility.

We would like to know when an integer is exactly divisible by (is a factor of) 2 through 12.

a. Most of you know that a number is divisible by 2 if it is even: it ends in a 2, 4, 6, 8, or 0.

b. Similarly, you know that if a number ends in 5 or 0, it is divisible by 5.

c. Also, if it ends in 0, it is divisible by 10.

d. If the sum of the digits is 3 or 9, the number is divisible by 3 or 9.

EXAMPLE 1—

Show that 771 is divisible by 3 but not 9.

7 + 7 + 1 = 15. 15 is divisible by 3 but not 9. 771/3 = 257, but 771/9 = 85% = 85⅔.

e. A number is divisible by 6 if the rule for 2 and 3 works.

f. A number is divisible by 4 if the last two digits are divisible by 4. Oh, let's show it!

EXAMPLE 2—

Show that 7,777,777,724 is divisible by 4.

7,777,777,724 = 7,777,777,700 + 24 = 77,777,777 × 100 + 24.

Since any number of hundreds is divisible by 4 and 24 is divisible by 4, so is the sum.

g. A number is divisible by 8 if the last three digits is divisible by 8.

NOTE

Divisible by 16 if the last 4 digits is divisible by 16, and so on.

h. A number is divisible by 12 if the rule for 4 and 3 holds.

NOTE

If you notice, 7 is missing. I have never learned the rule for 7, since dividing by 7 is simpler than the rule itself. These rules are supposed to make it easier, not harder, for you. These rules help reduce numerical fractions and occasionally show up on the SAT.

Rational Numbers

We have defined rational numbers in two ways, and we have shown how to change, let's say, 5/6 to a decimal by dividing 6 into 5.0000 and getting .8333 . . . = .8̄3̄. Let us now show how to change a repeating decimal into a fraction.

EXAMPLE 1—

Change .123123123 . . . to a fraction. Let
x = .123123123. . . .

Multiply both sides by 1 followed by the number of zeros of the repeating part of the decimal.

1000x = 123.123123123 . . .

$-$ x =   .123123123 . . .

999x = 123

$$x = \frac{123}{999} = \frac{41}{333}$$

EXAMPLE 2—

Change 3456.7464646 . . . to a fraction.

The repeating part is 46—multiply by 100.

100x = 345,674.646464 . . .

   x =   3,456.74646 . . .

99x = 342,217.9

$$x = \frac{342,217.9}{99} = \frac{3,422,179}{990}$$

The rational numbers are *dense.* This means that between any two different rational numbers is another rational. (The average, the arithmetic mean, is always between: add the two up and divide by 2). Irrationals and reals are dense, but integers are not dense because between 2 and 3 there is *not* another integer.

Irrational Numbers

Let us show that $\sqrt{2}$ is irrational. There are only two possibilities, $\sqrt{2}$ is rational or $\sqrt{2}$ is irrational. We will assume that $\sqrt{2}$ is rational and prove this wrong. The only conclusion is that $\sqrt{2}$ is irrational. (This is an *indirect proof.*) Suppose $\sqrt{2}$ is rational. Let $\sqrt{2} = a/b$, a, b integers in lowest terms (reduced). Then $2 = a^2/b^2$. $a^2 = 2b^2$. So a^2 is a multiple of 2. So a is a multiple of 2. $a = 2x$, x integer. Substitute. $a^2 = 2b^2$ gives $(2x)^2 = 2b^2$. Simplifying, we get $b^2 = 2x^2$. By exactly the same reasoning, we get b is a multiple of 2, so $b = 2y$, y integer. Now $a/b = 2x/2y$, which is impossible because a/b is in lowest terms, and now we can cancel the 2s. So $\sqrt{2} =$ rational is wrong. So $\sqrt{2}$ is irrational. (The proof is not easy—try to understand it.)

NOTE

With irrational numbers you can actually factor $x^2 - 7$.

Its factors are $(x + \sqrt{7})(x - \sqrt{7})$.

Also note: In a weird example, you can factor $x - 11$.

Its factors are $(\sqrt{x} + \sqrt{11})(\sqrt{x} - \sqrt{11})$!!!

You probably won't see such a problem, but it is interesting.

Pythagorean Theorem

Given a right triangle. Side c is the hypotenuse. The legs are a and b. The Pythagorean theorem states:

$$c^2 = a^2 + b^2$$

$$(\text{hypotenuse})^2 = (\text{leg}_1)^2 + (\text{leg}_2)^2$$

This is probably the most famous theorem (proven statement) in all mathematics. There are over 100 different proofs. Let us show one.

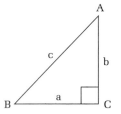

The area of the big square equals the area of the smaller square plus 4 congruent triangles.

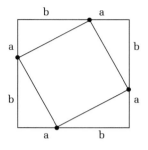

$(a + b)^2 = c^2 + 4(\frac{1}{2}ab)$

$a^2 + 2ab + b^2 = c^2 + 2ab$

Subtracting the 2ab from both sides, we get

$a^2 + b^2 = c^2$

(The proof isn't quite this easy because we actually have to show that the triangles are congruent [exactly the same] and that the smaller figure really is a square. If you've had plane geometry, you may be able to do this.)

Let's do some problems. There really are only two problems.

$5^2 + 6^2 = x^2$

$\quad x^2 = 61$

$\quad x = \sqrt{61}$

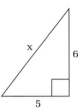

$x^2 + 7^2 = 9^2 \qquad$ hypotenuse always alone

$\quad x^2 = 81 - 49$

$\quad x = \sqrt{32}$

$\quad = \sqrt{(2)(2)(2)(2)(2)} = 4\sqrt{2}$

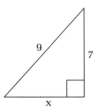

There are certain *pythagorean triples* that show up a lot later on. You should *memorize* these groups.

3, 4, 5 group　　(3² + 4² = 5²), largest number is always the hypotenuse

3, 4, 5 group　　3, 4, 5 . . . 6, 8, 10, . . . 9, 12, 15, . . . 12, 16, 20, . . . 15, 20, 25, . . .

5, 12, 13 group　　5, 12, 13, . . . 10, 24, 26, . . .

Aand 8, 15, 17, . . . 7, 24, 25, . . . 9, 40, 41, . . . 11, 60, 61,

. . . and my favorite 20, 21, 29 (only triple two digit where the first digit is the same.) I know I'm a dork.

There are two other special triangles: 45°-45°-90° and 30°-60°-90° right triangles. The SAT loves them. They are also very important later on!!!!

The 45-45-90 triangle comes from a square. Draw the diagonal.

$s^2 + s^2 = d^2$. Sooo, $d^2 = 2s^2$ and $d = s\sqrt{2}$.

Translating:

1. The legs are equal.

2. Leg to hypotenuse, multiply by $\sqrt{2}$.

3. Hypotenuse to leg, divide by $\sqrt{2}$.

Legs are equal. x = 7. Leg $(\sqrt{2})$ = hypotenuse. $y = 7\sqrt{2}$.

Hypotenuse is 10. Legs = hypotenuse/$\sqrt{2}$.

$$\text{So } x = y = \frac{10}{\sqrt{2}} = \frac{10 \cdot \sqrt{2}}{\sqrt{2} \cdot \sqrt{2}} = \frac{10\sqrt{2}}{2} = 5\sqrt{2}.$$

The 30-60-90 triangle comes from an equilateral triangle. All sides are equal. All angles = 60°. If you draw the altitude (the height), using geometry you can show that both triangles are congruent. So the short leg is half the side.

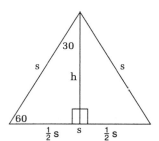

And the smallest angle is 30 = 180 − 90 − 60. Let us find h.

$$\left(\frac{1}{2}s\right)^2 + h^2 = s^2$$

$$\frac{1}{4}s^2 + h^2 = 1s^2$$

$$h^2 = \left(\frac{3}{4}\right)s^2$$

$$h = \frac{s\sqrt{3}}{2}$$

Again, let us translate:

   a. Short leg is opposite 30° angle.

   b. Long leg is opposite 60° angle.

   c. Hypotenuse is opposite 90° angle, as always.

   1. Always get short leg first.

   2. Hypotenuse to short: divide by 2.

   3. Short to hypotenuse: multiply by 2.

   4. Long to short: divide by $\sqrt{3}$.

   5. Short to long: multiply by $\sqrt{3}$.

   6. Also notice, you get the area of an equilateral triangle, which you will need later:

$$A = \frac{1}{2}bh = \frac{1}{2}(s)(s)\sqrt{3/2} = \frac{s^2\sqrt{3}}{4}$$

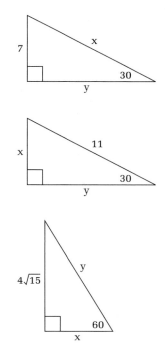

Short leg is given, 7. Short to hypotenuse, multiply by 2. x = 14. Short to long, multiply by $\sqrt{3}$. y = $7\sqrt{3}$.

Hypotenuse is given, 11. Hypotenuse to short, divide by 2. x = 5.5. Short to long, multiply by $\sqrt{3}$. y = $5.5\sqrt{3}$.

Long is given, $4\sqrt{15}$. Long to short, divide by $\sqrt{3}$. x = $4\sqrt{5}$. Short to hypotenuse, multiply by 2. y = $8\sqrt{5}$.

This should be easy, but some people have trouble and it is not always the poor students who are struggling!!!!

Real Numbers

We would like to mention properties of real numbers. Before we do that, we take a tiny side trip. We introduce the term *set*. It has no definition because it is an undefined word. A *set* is a collection of things. Braces, { }, indicate a set. The "things" in a set are called *elements*. In set theory, small letters are elements and capital letters are sets. {c, d} is a set with two elements in it. {c, d} = {d, c}. Order in a set doesn't matter. {c, d} = {c, c, c, d, d, c, c, d}. Different elements are counted once each. If we want to say, "4 is an element of the set {3, 5, 4}," we write $4 \in \{3, 5, 4\}$. $\in$ is the letter epsilon of the Greek alphabet, and is read, "is an element of."

The set of real numbers is a *field*. A field is a set S, with two operations, + and ×, which has the following properties:

Closed under + and ×: If a and b $\in$ S, then a + b and a × b $\in$ S. (If you add or multiply two members of the set, the answers are in the set, always.)

Associative: (a + b) + c = a + (b + c) and (a × b) × c = a × (b × c) for all a, b, c $\in$ S.

Commutative: a + b = b + a and a × b = b × a for all a, b $\in$ S.

Identity: For +, for all a $\in$ S, there is an element, call it 0, such that a + 0 = 0 + a = a.
For ×, for all a $\in$ S, there is an element, call it 1, such that a × 1 = 1 × a = a.

Inverse: For +, for all a $\in$ S, there is an element, call it −a, such that a + (−a) = (−a) + a = 0.
For ×, for all a $\in$ S, a ≠ 0, there is an element, call it (1/a), such that a × 1/a = 1/a × a = 1.

Distributive law: a × (b + c) = a × b + a × c.

The rationals also form a field but the irrationals do not.

For completeness, we state that the real numbers are *complete.* This means that each real number corresponds to a point on the real line, and every point on the line corresponds to a real number.

Complex (Imaginary) Numbers

Once upon a time, in the dream world of Math, Danny dreamed that there were more than natural numbers.

"I want the equation $x + 3 = 3$ to have an answer." And so the number 0 was born. This made Danny very happy. However, the next night Danny dreamed again.

"I want the equation $x + 9 = 2$ to have a solution." And so the rest of the integers, the negative integers, were born. Danny was very happy for a week. But Danny dreamed again.

"I want the equation $-5x = 3$ to have an answer." And so the rational numbers were born. Danny was happy for two weeks with this wonderful toy. But then Danny dreamed again.

"I want the equation $x^2 = 7$ to have an answer." And so the rest of the real numbers, the irrationals, were born. This made Danny happy for a whole month. But Danny dreamed again.

"I want the equation $x^2 = -1$ to have a solution." Danny thought and thought, but Danny knew there was no real solution to this problem since $(1)^2 = +1$ and $(-1)^2 = +1$.

"Aha. If there is no real number to answer the question, the answer must be imaginary!" So imaginary (and complex) numbers were born.

Okay, let's get back to reality.

DEFINITION

$\sqrt{-1} = i$. So, $i^2 = -1$.

EXAMPLE I

Write $\sqrt{-25}$ as an imaginary number.

$\sqrt{-25} = \sqrt{25} \times \sqrt{-1} = 5i$

DEFINITION

Complex numbers: Any number of the form $a + bi$, where a and b are reals and $i = \sqrt{-1}$.

If $a = 0$, the number is called *pure imaginary*. If $b = 0$, the number is a *real* number.

EXAMPLE 2

Write $\sqrt{16} - \sqrt{-36}$ as a complex number.

ANSWER

$4 - 6i$

NOTE

½ − ¼i and 2 + 3i are also complex numbers.

Here is a picture relating all of the kinds of numbers we have studied.

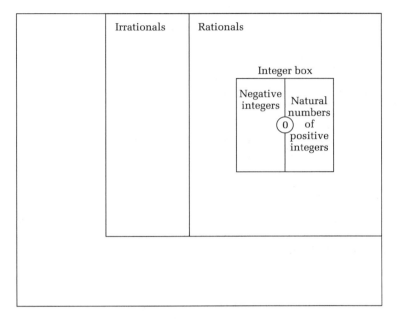

The *whole box* is the complex numbers.

The *large upper right box* is all the *real numbers,* and it is split into two parts: rationals and irrationals.

In reality, the set of complex numbers that are not real is lots bigger than the set of reals, and the set of irrationals is lots bigger than the set of rationals. The picture is drawn this way to easily show how the sets are related.

Now that we have complex numbers, like all other numbers, we would like to add them, subtract them, multiply them, divide them, and use powers. But first we'd like to know when complex numbers are equal.

Two complex numbers are equal if their real parts are equal and their imaginary parts are equal.

TRANSLATION

$a + bi = c + di$ if $a = c$ and $b = d$

EXAMPLE 3—

(x + 6) + (3y − 7)i = 2 + 8i. Find x and y.

x + 6 = 2. So x = −4. 3y − 7 = 8. So y = 5.

EXAMPLE 4—

Add (4 + 5i) + (7 + 3i).

I'll bet you can figure this one out yourself. Answer: 11 + 8i. Add real parts and add the imaginary parts; just like like terms in the first chapter.

EXAMPLE 5—

(4 + 5i) − (7 + 3i) = −3 + 2i

EXAMPLE 6—

Multiply (4 + 5i)(7 + 3i). How do we do this? Right, we FOIL it.

$28 + 12i + 35i + 15i^2$

$= 28 + 47i + 15(−1)$

$= 13 + 47i$

EXAMPLE 7—

Divide

$$\frac{(4 + 5i)}{(7 + 3i)}$$

and write the answer in a + bi form. How do we do this? Think of square roots. This was a tougher guess. Multiply top and bottom by the conjugate of the bottom. Just like with square roots, the *conjugate* of a + bi is a − bi (we change the sign of the imaginary part).

$$\frac{(4 + 5i) \times (7 − 3i)}{(7 + 3i) \times (7 − 3i)} = \frac{28 − 12i + 35i − 15i^2}{49 − 21i + 21i − 9i^2}$$

$$= \frac{28 + 23i - 15(-1)}{49 - 9(-1)} = \frac{43 + 23i}{58} = \frac{43}{58} + \frac{23}{58} i$$

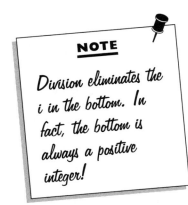

NOTE

Division eliminates the i in the bottom. In fact, the bottom is always a positive integer!

EXAMPLE 8—

Find i^{4567}. Oh, this is not as bad as it looks. We make the following observation: $i = \sqrt{-1} = i$, $i^2 = -1$, $i^3 = -i$, $i^4 = i^2 \times i^2 = (-1)(-1) = 1$, $i^5 = i$, and so on. Powers of i are *cyclic*—they repeat in cycles of 4. So the way to tell i^{4567} is very easy. Divide 4 into 4567. We don't care what the answer is, just the remainder R. If R = 1, the answer is i. If R = 2, the answer is −1. If R = 3, the answer is −i. If R = 0 (can't be 4), the answer is 1. 4567/4 = 1141 R3. The answer is −i.

EXAMPLE 9—

Suppose we want to graph the complex number −3 + 7i.

You cannot graph complex numbers on a real graph.

You graph complex numbers on a complex graph.

Instead of an x and a y axis, there is an x and a yi axis.

The picture looks like this:

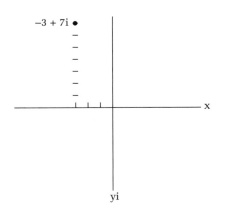

DEFINITION

The distance from point a + bi to the origin, 0 + 0i, is $|a + bi| = \sqrt{a^2 + b^2}$.

EXAMPLE 10—

Find the $|-3 + 7i|$

$$\sqrt{(-3)^2 + 7^2} = \sqrt{58}$$

NOTE

With imaginary numbers you can factor $a^2 + b^2$.

Its factors are (a + bi)(a − bi). To check, we multiply out and get $a^2 + abi − abi − b^2i^2 = a^2 − b^2i^2 = a^2 − (b^2)(-1) = a^2 + b^2$. So $x^2 + 9$ factors into (x + 3i)(x − 3i).

This is really all you need to know about complex numbers, with perhaps one minor optional exception a little later on, until past calculus. Complex numbers at this level are really not very complex.

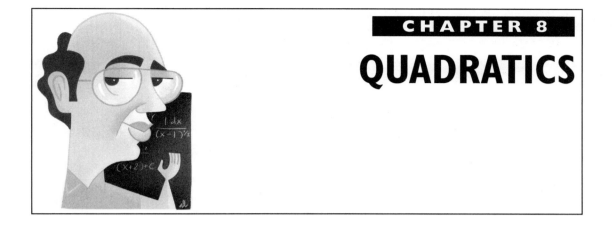

CHAPTER 8

QUADRATICS

We have previously solved quadratic equations, $ax^2 + bx + c = 0$, $a \neq 0$, by factoring. What happens if the quadratics don't factor or the roots are not rational? Let's do it!!

SOLVE BY SQUARE ROOTING

EXAMPLE 1

Solve for x: $x^2 - 7 = 0$.

$$x^2 = 7.$$

So $x = \pm\sqrt{7}$. The two roots are $+\sqrt{7}$ and $-\sqrt{7}$.

NOTE

This could be solved by factoring!! $x^2 - 7 = 0$. Difference of two squares. $(x + \sqrt{7})(x - \sqrt{7}) = 0$.

EXAMPLE 2

Solve for x:

$$a^2x^2 - b = 0 \qquad a^2x^2 = b$$

$$x^2 = \frac{b}{a^2}$$

$$x = \pm\sqrt{\frac{b}{a^2}} = \pm\frac{\sqrt{b}}{a}$$

EXAMPLE 3—

Solve for x:

$(x - 6)^2 - 2 = 0$

$(x - 6)^2 = 2$

$x - 6 = \pm\sqrt{2}$

$x = 6 \pm\sqrt{2}$

The two roots are $6 + \sqrt{2}$ and $6 - \sqrt{2}$.

SOLVE BY COMPLETING THE SQUARE

We need to have a perfect square. Let's see what one looks like.

If we square $(x + k)$, $(x + k)^2 = x^2 + 2kx + k^2$, we see that if the coefficient of x^2 is 1, half the coefficient of x squared (½ of 2k is k, . . . squared) will give us a perfect square.

EXAMPLE 1—

Complete the square $x^2 + 6x$.

$(6/2)^2$, half the coefficient of x, squared.

$x^2 + 6x + (6/2)^2 = x^2 + 6x + 9 = (x + 3)(x + 3) = (x + 3)^2$, a perfect square.

EXAMPLE 2—

Complete the square $x^2 - 9x$.

Coefficient of x is –9. Half is –9/2.

$x^2 - 9x + (-9/2)^2 = (x - 9/2)(x - 9/2) = (x - 9/2)^2$, again a perfect square.

We could now solve quadratics by completing the square.

EXAMPLE 3—

Solve $2x^2 - 7x + 3 = 0$ by completing the square.

1. $2x^2 - 7x + 3 = 0$

    1. Divide by the coefficient of x^2, in this case 2.

2. $x^2 - \dfrac{7}{2}x + \dfrac{3}{2} = 0$

    2. Bring the number term to the other side.

3. $x^2 - \dfrac{7}{2}x \qquad = -\dfrac{3}{2}$

    3. Complete the square. Take ½ of $-7/2 = -7/4$, square it, and add it to both sides.

4. $x^2 - \dfrac{7}{2}x + \left(-\dfrac{7}{4}\right)^2 = \left(-\dfrac{7}{4}\right)^2 - \dfrac{3}{2}$

    4. Factor the left; simplify the right.

5. $\left(x - \dfrac{7}{4}\right)^2 = \dfrac{49}{16} - \dfrac{3}{2} = \dfrac{49}{16} - \dfrac{24}{16} = \dfrac{25}{16}$

    5. Square-root both sides.

6. $\left(x - \dfrac{7}{4}\right) = \dfrac{\pm\sqrt{25}}{\sqrt{16}} = \dfrac{\pm 5}{4}$

    6. Bring the number to the right.

$$x = \dfrac{7}{4} \pm \dfrac{5}{4}$$

7. $x_1 = \dfrac{7}{4} + \dfrac{5}{4} \qquad x_2 = \dfrac{7}{4} - \dfrac{5}{4}$

    7. Get the two roots.

The two roots are x = 3 and x = ½.

Of course if given a choice, this one should be solved by factoring. Always factor if you have a choice and it can factor. Let's do one that can't factor, one with numbers, one with letters.

EXAMPLE 4—

Solve both for x.

1. $3x^2 - 9x + 2 = 0$

2. $x^2 - 3x + \dfrac{2}{3} = 0$

3. $x^2 - 3x = -\dfrac{2}{3}$

4. $x^2 - 3x + \left(-\dfrac{3}{2}\right)^2 = \left(-\dfrac{3}{2}\right)^2 - \dfrac{2}{3}$

5. $\left(x - \dfrac{3}{2}\right)^2 = \dfrac{9}{4} - \dfrac{2}{3}$

6. $\left(x - \dfrac{3}{2}\right)^2 = \dfrac{19}{12}$

7. $\left(x - \dfrac{3}{2}\right) = \dfrac{\pm\sqrt{19}}{\sqrt{12}} = \dfrac{\pm\sqrt{19}(\sqrt{3})}{2\sqrt{3}(\sqrt{3})}$

$= \dfrac{\pm\sqrt{57}}{6}$

8. $x = \dfrac{3}{2} \pm \dfrac{\sqrt{57}}{6}$

9. $x_1 = \dfrac{9 + \sqrt{57}}{6} \quad x_2 = \dfrac{9 - \sqrt{57}}{6}$

1. $ax^2 + bx + c = 0, a \neq 0$

2. $x^2 + \left(\dfrac{b}{a}\right)x + \dfrac{c}{a} = 0$

3. $x^2 + \left(\dfrac{b}{a}\right)x = -\dfrac{c}{a}$

4. $x^2 + \left(\dfrac{b}{a}\right)x + \left(\dfrac{b}{2a}\right)^2 = \left(\dfrac{b}{2a}\right)^2 - \dfrac{c}{a}$

5. $\left(x + \dfrac{b}{2a}\right)^2 = \dfrac{b^2}{4a^2} - \dfrac{c \times 4a}{a \times 4a}$

6. $\left(x + \dfrac{b}{2a}\right)^2 = \dfrac{b^2 - 4ac}{4a^2}$

7. $\left(x + \dfrac{b}{2a}\right) = \dfrac{\pm\sqrt{b^2 - 4ac}}{\sqrt{4a^2}}$

$= \dfrac{\pm\sqrt{b^2 - 4ac}}{2a}$

8. $x = \dfrac{-b}{2a} \pm \dfrac{\sqrt{b^2 - 4ac}}{2a}$

9. $x_1 = \dfrac{-b + \sqrt{b^2 - 4ac}}{2a} \quad x_2 = \dfrac{-b - \sqrt{b^2 - 4ac}}{2a}$

It is a good practice exercise, but boy is it long! The right-hand column is the shorter way to do it (just the answer!!!). It is called the *quadratic formula*.

THE QUADRATIC FORMULA

If $ax^2 + bx + c = 0$, $a \neq 0$, then

$$x = \frac{-b \pm \sqrt{b^2 - 4ac}}{2a}$$

are the two roots to this equation.

NOTE

a is the coefficient of x^2; b is the coefficient of x; c has no x.

EXAMPLE 1—

Solve for x, using the quadratic formula: $3x^2 - 9x + 2 = 0$.

$a = 3$ $b = -9$ $c = 2$

$$x = \frac{-b \pm \sqrt{b^2 - 4ac}}{2a} = \frac{-(-9) \pm \sqrt{(-9)^2 - 4(3)(2)}}{2(3)}$$

$$= \frac{9 \pm \sqrt{57}}{6}$$

Just like the last problem, only faster.

EXAMPLE 2—

Solve for x, using the quadratic formula: $4x^2 + 12x + 9 = 0$.

$a = 4$ $b = 12$ $c = 9$

$$x = \frac{-b \pm \sqrt{b^2 - 4ac}}{2a} = \frac{-12 \pm \sqrt{12^2 - 4(4)(9)}}{2(4)}$$

$$= \frac{-12 \pm \sqrt{144 - 144}}{2(4)} = \frac{-12 \pm 0}{8} = -\frac{3}{2}$$

The equation has two equal roots, both $-3/2$.

EXAMPLE 3—

Solve for x, using the quadratic formula: $x^2 - x + 8 = 0$.

$$a = 1 \qquad b = -1 \qquad c = 8$$

$$x = \frac{-b \pm \sqrt{b^2 - 4ac}}{2a} = \frac{-(-1) \pm \sqrt{1^2 - 4(1)(8)}}{2(1)}$$

$$= \frac{1 \pm \sqrt{-31}}{2}$$

There are two complex roots to this equation:

$$x_1 = \frac{1 + i\sqrt{31}}{2} \qquad \text{and} \qquad x_2 = \frac{1 - i\sqrt{31}}{2}$$

MORE ABOUT THE QUADRATIC FORMULA AND QUADRATIC EQUATION

Sometimes we don't need to know exactly what the solutions are but rather need to know the nature of the solutions—that is, are they real or not real, equal or not equal. To do that, we need to look what occurs under the square root sign. It has a name.

DEFINITION

If $ax^2 + bx + c = 0$, $a \neq 0$, then the *discriminant* is $b^2 - 4ac$. We assume a, b, c are real numbers.

1. If $b^2 - 4ac$ is negative, then the roots will be two complex conjugates.

EXAMPLE 1—

$x^2 - x + 8 = 0$. $a = 1$, $b = -1$, $c = 8$. $b^2 - 4ac$ $= (-1)^2 - 4(1)(8) = -31$.

The roots will be 2 complex conjugates. But this is just Example 3 from the preceding section!!!

2. If $b^2 - 4ac = 0$, there are two, equal, real roots.

EXAMPLE 2—

$4x^2 + 12x + 9 = 0$. $a = 4$, $b = 12$, $c = 9$. $b^2 - 4ac = 12^2 - 4(4)(9) = 0$.

There will be two, real, equal roots. But this is just Example 2 from the preceding section.

3. If $b^2 - 4ac$ is positive, there are two, real, unequal roots.

EXAMPLE 3—

$3x^2 - 9x + 2 = 0$. $a = 3$, $b = -9$, $c = 2$. $b^2 - 4ac = (-9)^2 - 4(3)(2) = 57$.

There will be two, real, unequal roots. But this is just Example 1 from the preceding section.

4. If $b^2 - 4ac$ is a perfect square, then the roots will be unequal and rational.

EXAMPLE 4—

$2x^2 - 7x + 3 = 0$. $a = 2$, $b = -7$, $c = 3$. $b^2 - 4ac = (-7)^2 - 4(2)(3) = 25$, a perfect square.

There are two, rational, unequal roots. But this is just Example 3 from "Solve by Completing the Square"!!!!

Given the quadratic equation $ax^2 + bx + c = 0$, we know the solutions, written a slightly different way are:

$$r_1 = \frac{-b}{2a} + \frac{\sqrt{b^2 - 4ac}}{2a} \quad \text{and} \quad r_2 = \frac{-b}{2a} - \frac{\sqrt{b^2 - 4ac}}{2a}$$

Suppose we add them together:

$$\frac{-b}{2a} + \frac{\sqrt{b^2 - 4ac}}{2a} + \frac{-b}{2a} - \frac{\sqrt{b^2 - 4ac}}{2a} = \frac{-2b}{2a} = \frac{-b}{a}$$

Suppose we FOIL them:

$$\left(\frac{-b}{2a} + \frac{\sqrt{b^2 - 4ac}}{2a}\right)\left(\frac{-b}{2a} + \frac{\sqrt{b^2 - 4ac}}{2a}\right)$$

$$= \frac{b^2}{4a^2} - \frac{b\sqrt{b^2 - 4ac}}{4a^2} + \frac{b\sqrt{b^2 - 4ac}}{4a^2} - \frac{(b^2 - 4ac)}{4a^2}$$

$$= \frac{b^2}{4a^2} - \frac{(b^2 - 4ac)}{4a^2} = \frac{b^2}{4a^2} - \frac{b^2}{4a^2} + \frac{4ac}{4a^2} = \frac{c}{a}$$

Hmmm.

To summarize:

If $ax^2 + bx + c = 0$, $a \neq 0$, then the sum of the roots will be $-b/a$ and the product of the roots will be c/a.

EXAMPLE 5—

$2x^2 - 7x + 3 = 0$. $a = 2$, $b = -7$, $c = 3$.

The sum of the roots is $-b/a = -(-7/2) = 7/2$.

The product of the roots is $c/a = 3/2$.

Again, this is "Solve by Completing the Square" Example 3. We found the roots were 3 and ½.

The sum is indeed 7/2 because $3 + \frac{1}{2} = 3\frac{1}{2} = 7/2$.

Aaaannnd the product is indeed 3/2 because $3(\frac{1}{2}) = (3/1)(1/2) = 3/2$!!!!

Finally, if we are given the roots can we find the equation? Yes, yes! Yes!!!! If r is a root, then $(x - r)$ is a factor.

EXAMPLE 6—

Suppose 4 and -6 are roots. Find an equation they came from.

If 4 is a root, $x - 4$ is a factor. If -6 is a root, then $x + 6$ is a factor. Soooo, $(x - 4)(x + 6) = 0$ and the equation is $x^2 + 2x - 24 = 0$.

EQUATIONS WITH RADICALS

Finally, we'd like to do equations involving square roots, partly because some of them result in quadratic equations, and partly because this is probably the only place to put these problems that won't totally disrupt the flow of the book.

When doing problems with square roots, we isolate (get alone on one side) the square root and square both sides.

EXAMPLE 1—

Solve for all values of x: $\sqrt{2x + 1} + 3 = 8$.

$\sqrt{2x + 1} + 3 = 8$ **Isolate the square root**

$\sqrt{2x + 1} = 5$ **Square both sides**

$2x + 1 = 25$ **Solve for x**

$2x = 24$

$x = 12$

To check: $\sqrt{2(12) + 1} + 3 \stackrel{?}{=} 8$. $5 + 3 = 8$. Yes, it checks. 12 is the root.

EXAMPLE 2—

Solve for all values of x: $\sqrt{4x} + 3 = 0$.

$\sqrt{4x} + 3 = 0$

$\sqrt{4x} = -3$

$4x = 9$

$x = \dfrac{9}{4}$

To check: $\sqrt{4(9/4)} + 3 \stackrel{?}{=} 0$. $3 + 3 \neq 0$. The equation has no solution.

> **NOTE**
>
> You must check. Some-times the answer(s) will check, sometimes not.

EXAMPLE 3—

Solve for all values of x: $\sqrt{2x + 1} + x = 7$.

Isolate the square root.

$$\sqrt{2x + 1} + x = 7$$

Square both sides.

$$\sqrt{2x + 1} = 7 - x$$

A quadratic. Get everything to one side.

$$2x + 1 = x^2 - 14x + 49$$

$$0 = x^2 - 16x + 48$$

Soooo x might be 12 or 4.

$$0 = (x - 12)(x - 4)$$

Check:

x = 12	x = 4
$\sqrt{2(12) + 1} + 12 \overset{?}{=} 7$	$\sqrt{2(4) + 1} + 4 \overset{?}{=} 7$
$5 + 12 \neq 7$	$3 + 4 = 7$

12 is not a solution!   4 is the only root.

NOTE

Sometimes both answers check, sometimes one answer checks, annnd sometimes none check!

ALSO NOTE

It is possible that you might have to use the quadratic formula, but almost nobody ever gives such a problem because the check would be awful.

EXAMPLE 4—

$\sqrt{5x + 1} - \sqrt{x + 6} = 1$. This is a long problem.

Isolate one of the square roots.

$$\sqrt{5x + 1} - \sqrt{x + 6} = 1$$

Square both sides; don't forget the messy middle term!!!!

$$\sqrt{5x + 1} = 1 + \sqrt{x + 6}$$

Collect like terms; isolate the remaining square root.

$$5x + 1 = 1 + 2\sqrt{x + 6} + x + 6$$

$4x - 6 = 2\sqrt{x + 6}$

Divide by the common factor if there is one. In this case, there is. Divide every term on both sides by 2.

$2x - 3 = \sqrt{x + 6}$

Again square both sides.

$4x^2 - 12x + 9 = x + 6$

$4x^2 - 13x + 3 = 0$

$(4x - 1)(x - 3) = 0$

$x = \dfrac{1}{4}$ or $x = 3$

Check:

$x = \dfrac{1}{4}$ $x = 3$

$\sqrt{5\left(\dfrac{1}{4}\right) + 1} - \sqrt{\dfrac{1}{4} + 6} \stackrel{?}{=} 1$ $\sqrt{5(3) + 1} - \sqrt{3 + 6} \stackrel{?}{=} 1$

$\sqrt{\dfrac{9}{4}} - \sqrt{\dfrac{25}{4}} \stackrel{?}{=} 1$ $\sqrt{16} - \sqrt{9} = 1$

$\dfrac{3}{2} - \dfrac{5}{2} \neq 1$

$\dfrac{1}{4}$ does not check.   3 checks.

POINTS, LINES, AND PLANES

POINTS

We would like to graph lines. Before we can do this, we must learn to graph points. First draw the horizontal real line. Mark it off in equal intervals, positive to the right, negative to the left. Call this the *x axis.* Draw another real line, perpendicular to the x axis, positive up, negative down. Call this the *y axis.* The graph is divided into four parts called *quadrants.* They are called I (because both x and y are positive), II (because, as we will see later, angles are measured counterclockwise positively), III, and IV as pictured. Points are given by *ordered pairs,* (x,y). The x number is always given first.

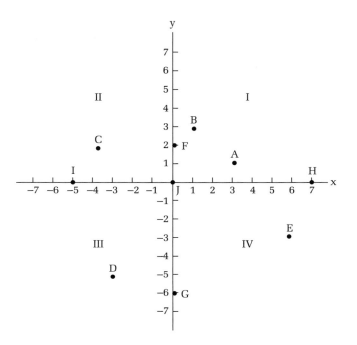

The x number is called the *abscissa* or *first coordinate*. The y number is called the *ordinate* or *second coordinate*. Together, (x,y) are the coordinates of the point.

Point A is the point (3,1), because it is 3 in the positive x direction and 1 in the positive y direction. It is not the same as B, which is (1,3), 1 in the x direction, 3 in the y direction. Ordered pairs mean the order matters!!!! C is the point (–4,2), 4 in the negative x direction, 2 in the positive y direction. D is (–3,–5). E is (6,–3), 6 in the positive x direction, 3 in the negative y direction. F is the point (0,2) and G is the point (0,–6). For *all points* on the y axis, the x coordinate is 0 because it is not to the left or to the right of the axis. H is (7,0) and I is (–5,0). For all points on the x axis, the y coordinate is 0 because the point is not above or below the axis.

Finally, where the axes (plural of *axis*) meet is the *origin,* point J. It is given by (0,0).

GRAPHING LINES

We would like very much to graph lines (when we use the word *line,* we mean infinite, straight). A line is of the form Ax + By = C, where A, B, and C are real numbers and A and B are both not zero. This is called *standard form.* We will talk about this a little later.

EXAMPLE 1

Graph the line y = 3x − 2.

1. Make a table.

2. Select values for x, usually three in case you make an error.

3. Substitute in the equation to get y.

x	y = 3x − 2	(x,y)
0	3(0) − 2 = −2	(0,−2)
1	3(1) − 2 = 1	(1,1)
3	3(3) − 2 = 7	(3,7)

4. Put the points on graph paper and connect the points. You should have a straight line.

The x values you choose don't have to be consecutive, but you should not take x = 100 since the point (100,298) is way off the paper!!! When you label the points, they all should be on one side of the line, annnd label the line at one end.

NOTE

This section is easier to do than to write. Practice graphing points. It shouldn't take too long to learn!

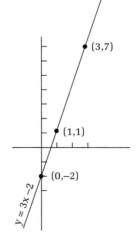

EXAMPLE 2—

Graph $2x + 3y = 12$.

First solve for y. Then do what you did in the last problem.

$2x + 3y = 12$

$3y = -2x + 12$

$y = \dfrac{-2x}{3} + \dfrac{12}{3}$

$y = \dfrac{-2x}{3} + 4$

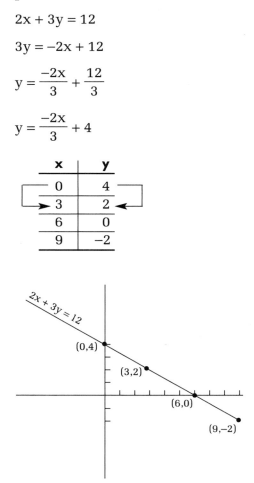

The x values we chose were multiples of 3. We did that so our arithmetic would be easier. We always want life as easy as possible. See what happens if you let $x = 1$!!!

Look at the arrows on the chart. From 4 to 2 is −2. From 0 to 3 is 3. The first over the second is −2/3, the *coefficient* of x (y = –⅔x + 4). This will happen every time. It has a name. It is called the *slope*. We will look at this in detail soon.

EXAMPLE 3—

Graph 2x + 3y = 12 the short way.

It is important to be able to graph lines the longer way, but if you do, the shorter way is . . . well, shorter. Because two points determine a line, we will use the easiest two. Please be careful because we use only two points here instead of three.

At the *x intercept,* the point where the line hits the x axis, the y coordinate = 0.

2x + 3y = 12

2x + 3(0) = 12

2x = 12

x = 6

We get the point (6,0). (x always goes first)

At the *y intercept,* the point where the line hits the y axis, x = 0.

2x + 3y = 12

2(0) + 3y = 12

3y = 12

y = 4

(0,4). Let's graph.

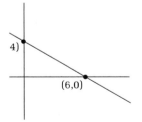

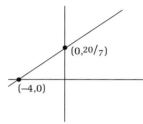

EXAMPLE 4—

Graph 5x − 7y = −20.

x intercept, y = 0. 5x = −20; x = −4; point is (−4,0).

y intercept, x = 0. −7y = −20; y = 20/7; point is (0,20/7).

Graph. With a little practice you may be able to do this way in your head.

This method of graphing is the easiest, but it doesn't work when there is only one intercept. When does this happen? Three cases: one parallel to the x axis; one parallel to the y axis; and one through the origin, not parallel to either axis.

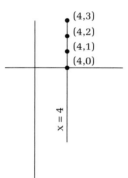

EXAMPLE 5—

Graph the points (4,0), (4,1), (4,3), (4,5).

You should notice two things about these points: the points form a vertical line and all the x values = 4. The equation of this line is x = 4.

NOTE

All vertical lines are x = something. The y axis is x = 0!!!!!!

EXAMPLE 6

If we graph points (0,3), (1,3), (5,3), you should notice that the points are horizontal and the y value is always 3. The equation of the line through these points is $y = 3$.

NOTE

All horizontal lines are y = something. The x axis is y = 0!!!!!!!

EXAMPLE 7

Graph $y = 3x$.

If we let x = 0, we get y = 0, the point (0,0). We can't let y = 0 because x = 0, the same point. So let x = something else—say, x = 2. So y = 6 and we have (2,6) and (0,0). Now graph!!!

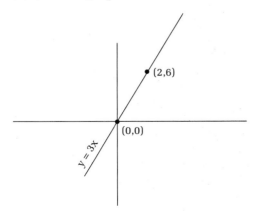

Before we proceed, it is important to know when we have a line!!!! Seems kind of silly, but let's make sure. We know a line is of the form $Ax + By = C$, where A, B are both not 0. It is important to understand this statement.

EXAMPLE 8—

For the following examples, tell whether the equations are straight lines, if they are of the form $Ax + By = C$. If they are, tell what A is, B is, and C is. If it is not a straight line, why not.

  a. $5x - 7y = -3$

  b. $5x + y = 0$

  c. $7x = 14$

  d. $y = -3x + 6$ Rewrite $3x + y = 6$

  e. $\dfrac{x}{6} - \dfrac{y}{3} = 1$

  f. $\dfrac{4}{x} + \dfrac{5}{y} = 7$

  g. $xy = 7$

  h. $x^2 + y = 7$

ANSWERS

  a. Yes. $A = 5$, $B = -7$, $C = -3$.

  b. Yes. $A = 5$, $B = 1$, $C = 0$.

  c. Yes!! $A = 7$, $B = 0$, $C = 14$. This is the vertical line $x = 2$! (In one dimension, this is a point.)

  d. Yes. $A = 3$, $B = 1$, $C = 6$.

  e. Yes. $A = 1/6$, $B = -1/3$, $C = 1$.

  f. No. x and y are in the denominator.

g. No. x and y are multiplied; they must be added or subtracted.

h. No. The variable must have exponent 1.

SLOPE

We would now like to talk about slope. We are given point P_1 with x coordinate x_1 and y coordinate y_1. x_1 and y_1 are numbers (constants). Draw the horizontal line and vertical line. Where they meet we'll call C. Every point on a horizontal line has the same height . . . so C = $(?,y_1)$. All points on a vertical line are the same distance from the x axis, in this case x_2 . . . . soooo C is (x_2,y_1). The length of P_2C: because the x values are the same, the length would be $y_2 - y_1$. For example, if C = (2,3) and P_2 = (2,8), the length of $P_2C = 8 - 3 = 5$, which is $y_2 - y_1$. Draw the picture and count the boxes. Similarly, $CP_1 = x_2 - x_1$ because the y's are the same.

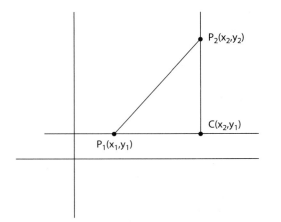

DEFINITION

$$slope = \frac{\text{change in y}}{\text{change in x}} = \frac{\Delta y}{\Delta x} = \frac{y_2 - y_1}{x_2 - x_1} = m$$

NOTE

The letter m is used for slope and Δ is the Greek letter delta, actually a capital delta, and is read, "change in."

EXAMPLE 1—

Find the slope of the line between (1,2) and (5,9).

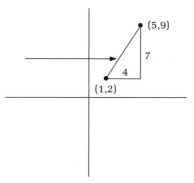

Let (1,2) be (x_1,y_1) and (5,9) be (x_2,y_2) (it could be the other way around).

$$m = \frac{y_2 - y_1}{x_2 - x_1} = \frac{9 - 2}{5 - 1} = \frac{7}{4}$$

If you go in the direction of the arrow and when you reach the line you have to go *up,* the slope is positive.

EXAMPLE 2—

Find the slope between (−3,4) and (2,−5).

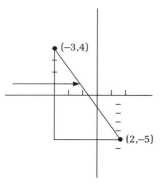

Let $(x_1,y_1) = (−3,4)$ and $(x_2,y_2) = (2,−5)$. So $x_1 = −3$, $y_1 = 4$, $x_2 = 2$, and $y_2 = −5$. (Careful of the minus signs.)

$$m = \frac{y_2 - y_1}{x_2 - x_1} = \frac{−5 − 4}{2 − (−3)} = \frac{−9}{5}$$

If you go in the direction of the arrow, and when you hit the line, you must go *down,* you have a negative slope.

EXAMPLE 3—

Find the slope between (1,3) and (7,3).

$$m = \frac{y_2 - y_1}{x_2 - x_1} = \frac{3 - 3}{7 - 1} = \frac{0}{6} = 0$$

Horizontal lines have slope = zero!!

EXAMPLE 4—

Find the slope between (1,3) and (1,8).

$$m = \frac{8 - 3}{1 - 1} = \frac{5}{0}$$

which is undefined.

Vertical lines have undefined slopes. In some books they call it *infinite slopes.*

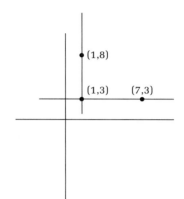

EQUATION OF THE LINE

Up to this point, if we wanted a line we were given the equation for it, except for vertical and horizontal lines. Now we are asked to find the line's equation, given some information. This topic is very important, not too long, and causes some students problems. Please read the next examples very carefully until you understand. Please don't worry if you don't understand. Most students do get this section with a little practice.

We have already learned standard form, that is, Ax + By = C, where A and B are both not zero together.

Point-slope form: Given a point (x_1, y_1) and the slope, m. Write the equation of the line.

Let (x,y) stand for any point on the line. By the definition of slope

$$m = \frac{\text{change in y}}{\text{change in x}} = \frac{y - y_1}{x - x_1}$$

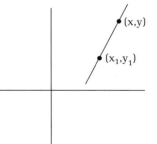

EXAMPLE 1

Find the equation of the line with point $(4,-7)$ and slope $3/11$.

$m = 3/11$ and $(x_1, y_1) = (4, -7)$. Sooo the equation is . . .

$$m = \frac{y - y_1}{x - x_1}$$

$$\frac{3}{11} = \frac{y - (-7)}{x - 4}$$

or $\quad \frac{3}{11} = \frac{y + 7}{x - 4}$

One more form is very important. It is called *slope intercept*. We will derive it from point slope.

Cross multiply.

$$\frac{m}{1} = \frac{y - y_1}{x - x_1}$$

Alternate version of point slope.

$$y - y_1 = m(x - x_1)$$

$$y - y_1 = mx - mx_1$$

$$y = mx + y_1 - mx_1$$

Now m is a number and x_1 is a number. So mx_1 is a number. y_1 is another number. Soooo, $y_1 - mx_1$ is a number. We give it a new name, b.

Slope intercept form: Given the slope m and y intercept b, the slope intercept form is $y = mx + b$.

Technically, the y intercept is $(0, b)$ because y intercept means the x coordinate is 0.

If you solve for y, the coefficient of x is the slope. This is very important!!

EXAMPLE 2

Find the equation of the line if the slope is $6/7$ and the y intercept is -3.

$m = 6/7$, $b = -3$. $y = mx + b$. Soooo, $y = (6/7)x - 3$.

Which form should you use? Most of the time, except for problems like Example 2, the easiest form is point slope. Let me try to convince you by doing a problem two different ways.

EXAMPLE 3—

Given points (4,7) and (9,15), find the equation of the line and write the answer in standard form.

No matter which method we use, we must find the slope.

$$m = \frac{15 - 7}{9 - 4} = \frac{8}{5}$$

Now we'll do it two different ways and see which is easier and shorter.

Method 1:

$$m = \frac{y - y_1}{x - x_1}$$

$$\frac{8}{5} = \frac{y - 7}{x - 4}$$ **Cross multiply.**

$$8(x - 4) = 5(y - 7)$$

$$8x - 32 = 5y - 35$$

$$8x - 5y = -3$$

Method 2:

$$y = mx + b$$

$$y = \frac{8}{5}x + b$$

$$(7) = \frac{8}{5}(4) + b$$

$$7 = \frac{32}{5} + b$$

$$\frac{35}{5} = \frac{32}{5} + b$$

$$b = \frac{3}{5}$$

Clear fractions

$$y = \frac{8}{5}x + \frac{3}{5}$$

$$5y = 8x + 3$$

**Standard form usually
has coefficient of x positive, although this is not
necessary.**

$$-8x + 5y = 3$$

Whew.

$$8x - 5y = -3$$

Why is the second method so much longer? It is because of all the fractions in the problem. If the slope was an integer, both methods would be about the same . . . buuuut the slope is not always an integer.* One more problem you should see.

EXAMPLE 4—

Find the equation of the line with x intercept 4 and y intercept −6.

x intercept means y = 0 . . . so the point is (4,0). y intercept means x = 0 . . . soooo the other point is (0,−6).

$$m = \frac{-6 - 0}{0 - 4} = \frac{-6}{-4} = \frac{3}{2}$$

So the equation of the line is

$$\frac{3}{2} = \frac{y - 0}{x - 4} \qquad \text{or} \qquad \frac{3}{2} = \frac{y + 6}{x - 0}$$

$$\text{or } y = \frac{3}{2}x - 6 \qquad \text{or} \qquad 3x - 2y = 12$$

whichever your teacher wants.

*We could have used the point (9,15). It will give the same answer. But we want the problem as easy as possible; so we use the easier point to do arithmetic.

A difficult twist of Example 4 is the following problem.

EXAMPLE 5—

Find the area of the triangle formed by the line through the point (4,7) and slope −2.

The first question is, "What triangle?" As you go on in math, *you must learn to draw pictures* to help you understand the problem.

First we draw the picture.

Suddenly we see the triangle, a right triangle. We also notice the base is the x intercept. The height is the y intercept. So the procedure is:

a. Find the equation of the line and write it in standard form.

b. Let y = 0 to find the x intercept, the base of the triangle.

c. Let x = 0 to find the y intercept, the height of the triangle.

d. Multiply one-half times the base times the height to find the area.

Using point slope,

$$m = \frac{y - y_1}{x - x_1}$$

In this case,

$$-2 = \frac{y - 7}{x - 4}$$

Writing −2 as −2/1 and cross multiplying, we get y − 7 = −2x + 8. In standard form, 2x + y = 15.

Letting y = 0, we get x = 15/2 = base.

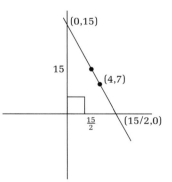

Letting $x = 0$, we get $y = 15 =$ height.

$$A = \tfrac{1}{2}bh = \tfrac{1}{2}\left(\frac{15}{2}\right)(15) = \frac{225}{4} = 56.25 \text{ square units}$$

Wow!!!

Fortunately, most of your problems will be a lot easier!

PARALLEL AND PERPENDICULAR LINES

Parallel lines means the slants are the same, so their slopes are the same.

Parallel: Let line L_1 have slope m_1 and line L_2 have slope m_2. Then L_1 is parallel to L_2 if $m_1 = m_2$ or if L_1 and L_2 are parallel to the y axis (and have no slope). Without proof, which will be shown later, we state the condition of perpendicular lines.

Perpendicular: L_1 is perpendicular to L_2 if $m_1 = -1/m_2$ (the slopes are negative reciprocals of each other or one line is parallel to the x axis and the other is parallel to the y axis).

EXAMPLE I

Let $y_1 = 3x - 8$; let $y_2 = -3x - 8$; let $y_3 = (1/3)x - 2$; and let $3x - y_4 = 6$. $y_5 = x - 6$.

Solving for y_4, we get $y_4 = 3x - 6$. Remember, if you solve for y, the coefficient of x is the slope $m_1 = 3$, $m_2 = -3$, $m_3 = 1/3$, $m_4 = 3$, aaand $m_5 = 1$. Sooooo y_1 and y_4 are parallel because $m_1 = m_4$. y_2 and y_3 are perpendicular because $m_2 = -1/m_3$. y_5 is not parallel or perpendicular to any of the lines.

EXAMPLE 2—

Let L = 5x + 6y = 9.

    a. Find the equation of the line parallel to L through the point (2,3).

    b. Find the equation of the line perpendicular to L through (4,−8).

In each case, we have to solve for y because the coefficient of x is the slope.

5x + 6y = 9

6y = −5x + 9

$$y = \frac{-5}{6}x + \frac{9}{6}$$

We don't care about 9/6 (the y intercept). The slope is −5/6.

ANSWERS

    a. Parallel means the same slope.

$$\frac{-5}{6} = \frac{y - 3}{x - 2}$$

    b. Perpendicular means the slope is the negative reciprocal.

$$\frac{+6}{5} = \frac{y + 8}{x - 4}$$

SOLVING TWO EQUATIONS IN TWO UNKNOWNS

We would like to find all the solutions common to two equations. There are a variety of ways to do it. We will do three in this chapter. All are pretty easy.

Solving by Graphing

Solving two equations in two unknowns by graphing is simplicity itself. Graph one line, then graph another line. Read where they meet.

EXAMPLE I—

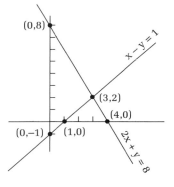

Solve for x and y by graphing:

$2x + y = 8$

$x - y = 1$

$2x + y = 8$. y intercept: $x = 0$, $y = 8$, point (0,8). x intercept: $y = 0$, $2x = 8$, $x = 4$, point (4,0).

Always do one line before you do the next.

$x - y = 1$. $x = 0$, $y = -1$, (0,−1). $y = 0$, $x = 1$, (1,0).

Using graph paper, we see the graphs meet at the point (3,2).

Let's check. $2x + y = 8$. $2(3) + 2 = 8$. $8 = 8$, okay. $x - y = 1$. $3 - 2 = 1$. Both check. The answer is $x = 3$ and $y = 2$.

Graphing is probably the least accurate way of doing this problem, especially if the answers are not integers. But sometimes, like in a chemistry or physics experiment, it is the only way to do the problem.

Solving by Substitution

Given two equations in two unknowns, we solve for one letter in one equation and substitute that letter in the second equation. We are allowed to do this because where the graphs meet, the x and y values are equal.

EXAMPLE 1—

Solve for x and y by substitution:

$2x + 3y = 5$

$x + 2y = 2$

We solve for the "easiest" letter, the x in the second equation, because the coefficient is 1.

Soooo

$x = 2 - 2y$

$2x + 3y = 5$

$2(2 - 2y) + 3y = 5$

$4 - 4y + 3y = 5$

$-y = 1$

So y = −1.

$x = 2 - 2y = 2 - 2(-1) = 4$

So the answer is x = 4, y = −1.

This method also has certain disadvantages. Let us show one.

EXAMPLE 2—

Solve for x and y:

$2x + 5y = 18$

$3x - 4y = 4$

There is no good letter to solve for. So let's solve for y in the first equation.

$$y = \frac{18 - 2x}{5}$$

$3x - 4y = 4$ becomes

$$3x - 4\left(\frac{18 - 2x}{5}\right) = 4$$

Multiplying by the LCD, 5, we get:

$15x -- 4(18 - 2x) = 20$

$\quad 15x - 72 + 8x = 20$

$\qquad\qquad 23x = 92$

$x = 4$

$$y = \frac{18 - 2x}{5} = \frac{18 - 2(4)}{5} = 2$$

The solution is $x = 4$ and $y = 2$. Again, it is messy because of the fractions. But doing it this way is necessary when we solve two equations in two unknowns where one or both are not straight lines. We'll do this in *PreCalc for the Clueless*.

In my opinion, the easiest method is . . .

Solving by Elimination

EXAMPLE I—

Solve for (x,y) by elimination:

$\quad x + 2y = \;\; 7$

$\underline{3x - 2y = 13}$

$\qquad 4x = 20$

$\qquad x = 5$

This is easy. We just add the equations. Substituting in either equation, we get $y = 1$ and the solution is (5,1).

EXAMPLE 2—

Solve for (x,y) by elimination:

$$4x + 3y = \quad 1$$

$$4x - 2y = \ 26$$

$$\overline{ 0 + 5y = -25}$$

$$y = -5$$

Sometimes we subtract!!!! Substituting, we get $x = 4$.
Answer: $(4,-5)$.

But sometimes . . . adding or subtracting doesn't work.

EXAMPLE 3—

Solve for (x,y) by elimination:

$$3x - 5y = 11$$

$$4x + 9y = -1$$

We will multiply the first equation by 4, and the second equation by -3. We then add the equations and the x's kill each other.

$$4(3x - 5y = 11) \qquad 12x - 20y = 44$$

$$-3(4x + 9y = -1) \qquad \underline{-12x - 27y = \ 3}$$

$$-47y = 47 \qquad y = -1$$

There are two ways to finish the problem:

Method 1: Substitution as before.

$$y = -1$$

$$3x - 5y = 11$$

$$3x - 5(-1) = 11$$

$$3x + 5 = 11$$

$$3x = 6$$

$$x = 2$$

The solution is $(2,-1)$.

But if the y value were very messy . . . use. . . .

Method 2: Double elimination. We eliminated x the first time. Now eliminate y. We multiply the first equation by 9 and the second by 5.

$$9(3x - 5y = 11) \qquad 27x - 45y = 99$$

$$5(4x + 9y) = -1 \qquad \underline{20x + 45y = -5}$$

$$47x \qquad = 94 \qquad x = 2$$

The solution is again (2,−1)

If the signs are different, both multiplications must be the same sign. I suggest two plus signs unless you want to make life really difficult for yourself.

Just for fun, let's do a messy one (still with a nice answer).

NOTE

In the elimination method, if the signs are the same in the letter you eliminate, when you multiply, one number must be plus and one minus.

EXAMPLE 4—

Solve for x and y:

Find separate LCD on each line and line up x and y terms. Top line by 6 and bottom by 8.

$$\frac{3}{2} x + \frac{4}{3} y = 20$$

$$\frac{1}{8} x + 2 = \frac{1}{2} y$$

Next multiply the top by 1 and the bottom by 2 and add to eliminate y.

$$9x + 8y = 120 \quad (1$$

$$x - 4y = -16 \quad (2$$

$$9x + 8y = 120$$

$$2x - 8y = -32$$

$$11x = 88$$

x = 8

Substituting, y = 6. The answer is (8,6).

Relating the Graphs to the Algebra

Well, what do all of these problems have in common?
Let's relate the two together.

EXAMPLE 1—

Algebra:

x + y = 6

x − y = 2

2x = 8

So x = 4. Substituting, y = 2. The answer is the point
(4,2).

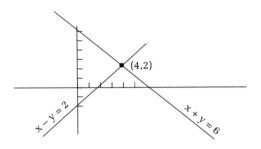

EXAMPLE 2—

Algebra:

−2)　　x + y = 2　　　−2x − 2y = −4

1)　2x + 2y = 8　　　2x + 2y = 8

Adding　　　　　　　　　　0 = 4

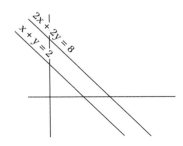

0 = 8? When does 0 = 8? Answer, *never!!!!* No solution
to the problem. Let's see what the graph looks like. . . .
The graphs never meet. No solution means the lines
are parallel.

EXAMPLE 3—

Algebra:

3) $x + 2y = 4$ $3x + 6y = 12$

−1) $3x + 6y = 12$ $-3x - 6y = -12$

$0 = 0$

When does $0 = 0$? *Always!!!* Let's see the graph.

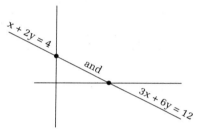

Both lines are the same. There is an infinite number of solutions, namely every point on the line $x + 2y = 4$.

In solving two equations in two unknowns, there are three possibilities: one point (x,y) is the answer, there are no solutions (parallel lines), or an infinite number of solutions (every point on a line) is the answer.

WORD PROBLEMS

Old Ones Revisited

Many times in math, when you learn new techniques, older problems that may at one point have been difficult become easier, much easier, or much, much, much easier. This is especially true about the word problems we had before. Let us look at a few we've seen before.

EXAMPLE 1—

The sum of two numbers is 42 and their difference is 10. Find them.

This one is much, much, much easier. Let x = larger number and y = smaller.

$x + y = 42$

$x - y = 10$

$2x = 52$

So x = 26. Sooo y = 16. That's it.

EXAMPLE 2—

How many pounds of Columbian coffee at $7 a pound must be mixed with $5 a pound Brazilian coffee to give us 14 pounds of the mixture at $5.50 a pound?

Let x = pounds of Columbian coffee.

Let y = pounds of Brazilian coffee. The table is . . .

	Cost/pound ×	pounds =	total cost
Columbian	7.00	x	7x
Brazilian	5.00	y	5y
Mixture	5.50	14	77

5) $x + y = 14$

−1) $7x + 5y = 77$

        $5x + 5y = 70$

        $-7x - 5y = -77$

        $-2x = -7$

          $x = 3.5$ pounds of Columbian coffee

This is easier in two unknowns, as you can see. The next one would be really bad in one unknown. Let's do it in two.

EXAMPLE 3—

8 years ago Jamie's age was ⅔ of Billy's age. In 4 years, Jamie's age will be ¾ of Billy's age. What are their ages today?

We let x = Jamie's age now and y = Billy's age now. Make the chart.

	Now	8 yrs ago	4 yrs from now
Jamie	x	x − 8	x + 4
Billy	y	y − 8	y + 4

8 years ago, Jamie's age was ⅔ Billy's age.

$$x - 8 = \left(\frac{2}{3}\right)(y - 8) \quad (3$$

In 4 years, Jamie will be ¾ Billy's age.

$$x + 4 = \left(\frac{3}{4}\right)(y + 4) \quad (4$$

$$3(x - 8) = 2(y - 8)$$

$$4(x + 4) = 3(y + 4)$$

$$3x - 24 = 2y - 16 \quad \text{or} \quad 3x - 2y = 8 \quad (4$$

$$4x + 16 = 3y + 12 \quad \text{or} \quad 4x - 3y = -4 \quad (-3$$

$$12x - 8y = 32$$

$$-12x + 9y = 12$$

y = 44, the age of Billy. By substitution, x = 32, the age of Jamie.

New Problems

EXAMPLE 1—

A fraction reduces to ⅔. If 4 is subtracted from the numerator and 6 is added to the denominator, the new fraction reduces to ⅖. Find the original fraction.

Let n = numerator (very clever letter) and d = denominator.

$\dfrac{n}{d} = \dfrac{2}{3}$ or $2d = 3n$ or $2d - 3n = 0$

$\dfrac{n - 4}{d + 6} = \dfrac{2}{5}$ or $2(d + 6) = 5(n - 4)$ or

$2d - 5n = -32$ Substitute 3n for 2d!   **Top minus bottom**

$-2n = -32$

$n = 16$

Sooo, d = 24.

This type is called a *digit problem.* Suppose we have a two-digit number 37. 37 = 30 + 7 = 10(3) + 7. If we represent a two-digit number by tu, where t = tens digit and u = units digit, the number is represented by 10t + u.

The number reversed is 73 = 7(10) + 3. Sooo ut = 10u + t.

7 + 3 is the sum of the digits. t + u is the sum of the digits in letters. Let's go.

EXAMPLE 2—

The sum of the digits of a two-digit number is 8. If the digits are reversed, the new number is 36 more than the old number. Find the original number.

$$t + u = 8 \qquad\qquad \text{or} \qquad t + \quad u = \quad 8 \quad (9$$
$$10u + t = 10t + u + 36 \qquad\qquad -9t + \quad 9u = \quad 36 \quad (1$$

$$9t + \quad 9u = \quad 72$$
$$-9t + \quad 9u = \quad 36$$
$$18u = 108$$
$$u = 6$$

Sooo t = 2. The number is 26. To check, 62 − 26 = 36.

A New Distance Problem

We row a boat at 6 miles per hour. The speed of the stream is 2 mph. Upstream we travel at 4 mph (6 − 2). Downstream, 8 mph (6 + 2).

EXAMPLE 1

A plane traveling 4 hours with the wind goes 1360 miles and traveling 5 hours into the wind goes 1300 miles. Find the speed of the plane and the speed of the wind.

Let x = speed of the plane annnnd y = speed of the wind. With the wind, speed is x + y. Against, issss x − y.

	R	**t**	**=**	**d**
With	x + y	4		1360
Against	x − y	5		1300

$$4(x + y) = 1{,}360$$
$$5(x - y) = 1{,}300$$

5) $\quad 4x + 4y = 1{,}360$

4) $\quad 5x - 5y = 1{,}300$

$$20x - 20y = 6{,}800$$
$$20x + 20y = 5{,}200$$
$$40x \qquad = 12{,}000$$

So x = 300 mph, the speed of the plane. Substituting, y = 40 mph, the speed of the wind.

THREE EQUATIONS IN THREE UNKNOWNS

Why We Are *Not* Doing This by Graphing

We want to solve three linear equations in three unknowns. But if we graph this, we find that we have to graph in three dimensions. Three-dimensional graphing is usually part of Calculus III in most colleges. We are going to do it a little here, but it is *just for fun*. Since many of you will be able to do this process, it should be clear that math does *not* go from easy to hard. Remember, this is just for fun!!!

EXAMPLE I—

Graph the point (3,4,5) in 3 dimensions.

The picture is as shown on page 177. The y and z axes are in the paper, positive z up, negative z down, positive y to the right, negative y to the left. x axis at an angle of approximately 45° to the y axis, positive out

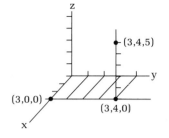

of the paper negative into the paper at 70 percent of the scale of the y and z axis scales, which are the same. All points are ordered triples (x,y,z).

Go 3 in the x direction; get the point (3,0,0). Go 4 in the y direction and we get point (3,4,0). 5 up gives the point (3,4,5). It helps to be a little bit of an artist. I'm sure some of you are better than me. Fortunately, we do not graph like this.

EXAMPLE 2—

Graph the plane $2x + 3y + 4z = 12$.

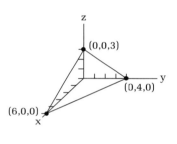

We will do this using intercepts. $x = 0$, $y = 0$, $z = 3$, (0,0,3). $x = 0$, $z = 0$, $y = 4$, (0,4,0). $y = 0$, $z = 0$, $x = 6$, (6,0,0). The graph looks like the picture.

EXAMPLE 3—

Graph the plane $5x + 2y = 10$.

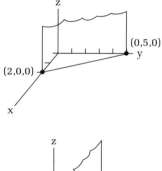

In two dimensions, this is a line. In 3D, this is a plane. $x = 0$, $y = 5$ and $y = 0$, $x = 2$ buuut z can be anything. Sooooo the graph looks like the picture.

EXAMPLE 4—

Graph the plane $y = 4$.

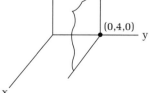

In one dimension, $y = 0$ is a point. In 2D, it is a line. In 3D, it is a plane. $y = 4$ but x and z can be anything. Soooo the picture is. . . . (Parallel to the xz plane, which would be $y = 0$.)

You can see how impossible it would be to solve three equations by graphing. First, they would all be on the same graph. Next, if you could do this, then you would have to see where the planes meet!! However, we can ask, "What would the solutions look like?"

1. The most common gives us a point, one x value, one y value, and one z value. If you intersect two planes you get a line. A line with the third plane gives a point.

2. You can get no points of intersection. One example is three parallel planes.

3. Like the binding of a book, the intersection of the planes could be a line.

4. The three planes could be the same planes, and the answer would be a plane.

NOTES

1. Case 1 will be the vast majority.

2. We will mostly use the elimination method, with occasionally some substitution.

Solving Three Equations in Three Unknowns

EXAMPLE 1—

Solve:

$$3x + y + 2z = 14 \quad (1)$$

$$2x - 2y + 5z = 14 \quad (2)$$

$$x - y + 6z = 14 \quad (3)$$

First, all the numbers on the right don't have to be the same. I'm just being cute! Next, we must reduce three equations in three unknowns to two equations in two unknowns by eliminating one of the letters. The easiest letter here to eliminate is . . . y. . . .

Adding Eqs. (1) and (3) eliminates y.

$$3x + y + 2z = 14$$

$$\underline{x - y + 6z = 14}$$

$$4x + + 8z = 28 \qquad (4)$$

Adding 2 times Eq. (1) and Eq. (2) also eliminates y.

$$2(3x + y + 2z = 14)$$

$$1(2x - 2y + 5z = 14)$$

$$6x + 2y + 4z = 28$$

$$\underline{2x - 2y + 5z = 14}$$

$$8x + 9z = 42 \qquad (5)$$

Let's do it.

Now we solve two equations in two unknowns as before:

$$-2) \quad 4x + 8z = 28 \qquad (4)$$

$$1) \quad 8x + 9z = 42 \qquad (5)$$

$$-8x - 16z = -56$$

$$\underline{8x + 9z = 42}$$

$$-7z = -14$$

$z = 2$. Substitute into one of the two equations in two unknowns [in (4) or (5)].

$$8x + 9z = 42$$

$$8x + 9(2) = 42$$

$$8x = 24$$

$$x = 3$$

Substitute x = 3 and z = 2 into one of the three equations in three unknowns [in (1) or (2) orrr (3)].

$$2x - 2y + 5z = 14$$

$$2(3) - 2y + 5(2) = 14$$

$$-2y = -2$$

$$y = 1$$

Whew! The answers are x = 3, y = 1, z = 2, or the point (3,1,2).

These problems are very long, but not hard. You must be very, very careful. But they're not as bad as four equations in four unknowns. More than this is done now on computers.

Let us do two more.

EXAMPLE 2—

Solve:

$$x + y + z = +1 \quad (1)$$

$$-2x + y + z = -2 \quad (2)$$

$$3x + 6y + 6z = 7 \quad (3)$$

If we take Eq. (1) − Eq. (2), something unusual happens. We get 3x = 3. x = 1. We then substitute x = 1 in Eqs. (1) and (3).

$$1 + y + z = 1$$

$$3 + 6y + 6z = 7$$

$$-6) \quad y + z = 0$$

$$1) \quad 6y + 6z = 4$$

$$-6y - 6z = 0$$

$$6y + 6z = 4$$

We get 0 = 4.

Because 0 never equals 4, there is no solution to this problem.

EXAMPLE 3—

Solve:

$$2x - y + z = -1 \quad (1)$$

$$x + 3y - 2z = 2 \quad (2)$$

$$-5x + 6y - 5z = 5 \quad (3)$$

Let's eliminate x.

1 times Eq. (1) plus −2 times Eq. (2).

$$1(2x - y + z) - 2(x + 3y - 2z) = 1(-1) - 2(2)$$

5 times Eq. (2) plus 1 times Eq. (3).

$$5(x + 3y - 2z) + (-5x + 6y - 5z) = 5(2) + 5$$

$$-7y + 5z = -5 \quad (3$$

$$21y - 15z = 15 \quad (1$$

$$-21y + 15z = -15$$

$$21y - 15z = 15$$

$$0 = 0$$

Because $0 = 0$, it means there is an infinite number of solutions. Here we will take it to mean that one letter can equal anything. Say $z = k$ (k is for constant because mathematicians can't spelll):

$-7y + 5z = -5 \qquad z = k$

Sooo $-7y + 5k = -5$

$-7y = -5k - 5$

So $y = \dfrac{(-5k - 5)}{(-7)} \qquad \text{or} \qquad \dfrac{5}{7}(k + 1)$

$z = k \qquad \text{and} \qquad y = \dfrac{5}{7}(k + 1)$

We can now find x. We know $x + 3y - 2z = 2$. Sooo . . .

$x + 3\dfrac{5}{7}(k + 1) - 2k = 2$

$x = 2k + 2 - \dfrac{15}{7}k - \dfrac{15}{7}$

$\quad = \dfrac{14k + 14 - 15k - 15}{7}$

$\quad = \dfrac{-k - 1}{7} = \dfrac{-1}{7}(k + 1)$

The answers are the infinite set of triples.

$$\left(\dfrac{-1}{7}(k + 1), \dfrac{5}{7}(k + 1), k\right)$$

where k is any real number. As we said before, most of the time we get one value for x, one for y, and one for z.

WORD PROBLEMS IN THREE DIMENSIONS

Finally, let's do a few word problems with three variables.

EXAMPLE 1—

We have nickels, dimes, and quarters, totaling $3.30, 21 coins in all, with one more nickel than dimes.

	Value of a coin $\times$ number of coins = total value		
Nickel	5	n	5n
Dime	10	d	10d
Quarter	25	q	25q
Mix		21	330

$$n + d + q = 21$$

$$5n + 10d + 25q = 330$$

Let's do part by substitution. Nickels equal one more than dimes, or n = d + 1. So

$$d + 1 + d + q = 21$$

$$5(d + 1) + 10d + 25q = 330$$

$$2d + q = 20 \qquad q = 20 - 2d$$

$$15d + 25q = 325$$

$$15d + 25(20 - 2d) = 325$$

$$-35d = -175$$

$$d = 5 \qquad n = 6 \qquad q = 10$$

Not all three equations in three unknowns are bad. Here's an even nicer one.

EXAMPLE 2—

A three-digit number has the sum of the digits 9. If the digits are reversed, the new number is 99 less than the original. Annnd the sum of the hundreds digit and tens digit is twice the units digit. Find the number.

For htu, the sum of the digits is h + t + u, the number is 100h + 10t + u, and the number reversed is 100u + 10t + h.

$$h + t + u = 9 \quad (1)$$

$$h + t = 2u$$

or $\qquad h + t - 2u = 0 \quad (2)$

$$100h + 10t + u - (100u + 10t + h) = 99$$

$$99h - 99u = 99$$

$$h - u = 1 \quad (3)$$

If we are clever, we divide by 99.

Something really nice happens here. Equation (1) − Eq. (2) gives $3u = 9$. $u = 3$. Putting 3 into Eq. (3), we get $h - 3 = 1$. So $h = 4$. Putting both into Eq. (1), we get $4 + t + 3 = 9$. So $t = 2$. The number is 423, with very little work.

We now go to sort of a mishmash chapter, a little of this and a little of that.

CHAPTER 10

ODDS AND ENDS

RATIO AND PROPORTION

A *ratio* is two items being compared to each other and written as a fraction.

EXAMPLE 1

The ratio of 3 to 7.

ANSWER

$\dfrac{3}{7}$

EXAMPLE 2

The ratio of 2 feet to 7 inches.

2 ft = 2 × 12 = 24 inches

ANSWER

$\dfrac{24}{7}$

Proportion: 2 ratios equal to each other.

EXAMPLE 3

Four is to five as 7 is to x.

ANSWER

$$\frac{4}{5} = \frac{7}{x}$$

Of course, we can solve proportions with unknowns in them. In this case, $4x = 35$. $x = 35/4$.

Similar triangles: Triangles with corresponding angles equal and corresponding sides in proportion.

In these pictures, corresponding angles are equal.

$\angle A = \angle X$, $\angle B = \angle Y$, $\angle C = \angle Z$.

Corresponding sides are in proportion.

$$\frac{a}{x} = \frac{b}{y} = \frac{c}{z}$$

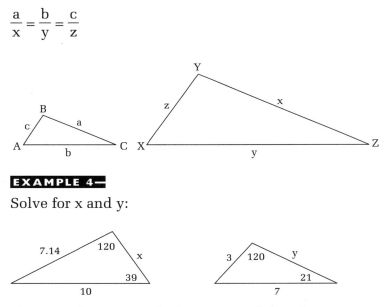

EXAMPLE 4—

Solve for x and y:

The triangles are similar because each has the same angles: $21°$, $39°$, and $120°$. Remember, the angles add up to $180°$. But the triangles are backward. Soooo 7 corresponds to 10, x corresponds to 3, and 7.14 corresponds to y. We will do 2 different proportions. Always start with the ratio where you have both numbers, if possible.

$$\frac{7}{10} = \frac{3}{x} \qquad 7x = 30 \qquad x = \frac{30}{7}$$

$$\frac{7}{10} = \frac{y}{7.14} \qquad 10y = 50 \text{ (approx)} \qquad y = 5$$

GEOMETRIC FORMULAS

This section is just to make sure you know certain additional geometry formulas.

Trapezoid: Quadrilateral with exactly one pair of parallel sides.

$p = b_1 + b_2 + j + k$

$A = \dfrac{1}{2} h(b_1 + b_2)$

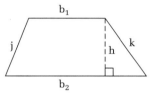

Cube: e = edge

Volume $e \times e \times e = e^3$ (Cubing comes from a cube!!!!)

Surface area $= 6e^2$ (6 squares)

Diagonal $= e\sqrt{3}$

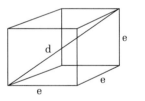

Box or, technically, a *rectangular parallelepiped.*

$V = l \times w \times h$

$SA = 2\,l \times w + 2\,w \times h + 2\,l \times h$ (6 sides—rectangles)
      top &   sides   front
      bottom   & back

$d = \sqrt{l^2 + w^2 + h^2}$ (3D Pythagorean theorem)

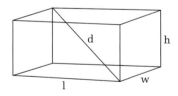

Sphere

$V = \dfrac{4}{3} \pi r^3$

$SA = 4\pi r^2$

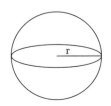

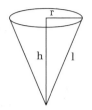

Cone

$$V = \frac{1}{3}\pi r^2 h$$

$$SA = \pi r^2 + \pi rl$$

     top   curved

$$l = \sqrt{r^2 + h^2}$$

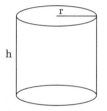

Cylinder

$$V = \pi r^2 h$$

$$SA = 2\pi r^2 + 2\pi rh$$

      top &   side
      bottom

Oh, let's do some problems with these formulas.

EXAMPLE 1—

For the cube, find the volume, surface area, and diagonal.

$$V = e^3 = 5^3 = 125 \text{ cubic units}$$

$$SA = 6e^2 = 6(5)^2 = 150 \text{ square units}$$

$$\text{Diagonal} = e\sqrt{3} = 5\sqrt{3}$$

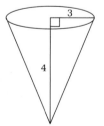

EXAMPLE 2—

Find the slant height, surface area, and volume of the cone.

$$l = \sqrt{3^2 + 4^2} = \sqrt{25} = 5 \text{ units} \qquad \text{A Pythagorean}$$
$$\text{triple again!!!!!)}$$

$$SA = \pi rl + \pi r^2$$

$$= \pi 3(5) + \pi 3^2 = 24\pi \text{ square units}$$

$$V = \left(\frac{1}{3}\right)\pi r^2 h = \left(\frac{1}{3}\right)\pi 3^2(4) = 12\pi \text{ cubic units}$$

If a 3-D solid comes to a point, the volume is multi-plied by ⅓. Another example is a pyramid. Oh, let's do a pyramid for fun.

EXAMPLE 3—

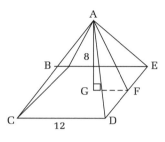

Square base, CD = 12. Height of pyramid = 8.

Find volume, slant height AF, surface area, edge AE.

$$V = \frac{1}{3} \quad s^2h = \left(\frac{1}{3}\right)12^2(8) = 384 \text{ cubic units}$$

AF, the slant height, is another Pythagorean triple: 6, 8 . . . AF = 10.

The surface area is the base, a square plus 4 equal triangles.

$$SA = s^2 + 4\left(\frac{1}{2}\right)bh = 12^2 + 4\left(\frac{1}{2}\right)(12)(10) = 384 \text{ square units}$$

(The base GF is 1/2 of CD. See??!!)

For the edge AE, you must see a right angle AFE. Soooo AF² + EF² = AE². EF is 1/2 of DE, which is also 6.

$$6^2 + 10^2 = AE^2$$

$$AE = \sqrt{136} = 2\sqrt{34} \text{ units}$$

This is a popular problem in geometry books. Let's go on. . . .

EXAMPLE 4—

Find the area and the perimeter of the trapezoid.

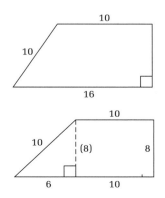

I decided to show you a slightly tricky one. The trick is to drop another height as pictured. We get a right trian-gle. The whole bottom base is 16. The part in the rect-angle is 10. Sooooo the base of the triangle is 6. The height and the right side (also the height) is 8 because we have a 6–8–10 Pythagorean right triangle. These things pop up all the time. Teachers love them because they don't have to do any arithmetic.

$p = 10 + 10 + 8 + 16 = 44$ units   (Other 8 is not part of the perimeter.)

$A = \dfrac{1}{2}\,h(b_1 + b_2) = \dfrac{1}{2}\,(8)(10 + 16) = 104$ square units

This is the kind of SAT problem that is harder than it should be because textbooks today do not have many (any?) of these problems.

EXAMPLE 5—

Let $OA = 10$. Find the area of the shaded region.

We see that the region is one-quarter of a circle minus a triangle.

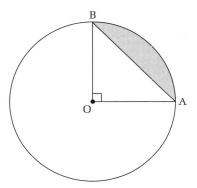

In formulas, it is $\frac{1}{4}\pi r^2 - \frac{1}{2}bh$. $r = OA = OB = b = h = 10$.

So $A = \frac{1}{4}\pi\,(10)^2 - \frac{1}{2}(10)(10) = 25\pi - 50$ square units.

EXAMPLE 6—

Find the volume, surface area, and diagonal.

$V = l \times w \times h. = 12 \times 3 \times 4 = 144$ cubic units

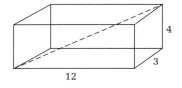

$$SA = 2(12)(3) + 2(3)(4) + 2(12)(4)$$
$$\underset{\text{bottom}}{\text{top \&}} \qquad \text{sides} \qquad \underset{\text{back}}{\text{front \&}}$$

$= 72 + 24 + 96 = 192$ square units

$d = \sqrt{l^2 + w^2 + h^2} = \sqrt{12^2 + 3^2 + 4^2} = \sqrt{169} = 13$ units

EXAMPLE 7—

Find the volume and surface area of a sphere with radius 6.

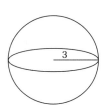

$$V = \left(\frac{4}{3}\right)\pi r^3 = \left(\frac{4}{3}\right)\pi(3)^3 = 36\pi \text{ cubic units}$$

$SA = 4\pi r^2 = 4\pi(3)^2 = 36\pi$ square units

And what is the significance of the same number? Absolutely none. It is just what it is.

EXAMPLE 8—

Find the volume and surface area of a right circular cylinder.

If d = 20, then r = 10.

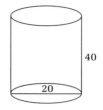

40

20

$V = \pi r^2 h = \pi 10^2 (40) = 4000\pi$ cubic units

$SA = 2\pi r^2 + 2\pi rh = 2\pi 10^2 + 2\pi(10)(40) = 1000\pi$ square units

We know a little about circles. Let's summarize a bunch of more facts about the circle. If you take a geometry class, you will probably have a bunch of these if not all of them.

We really can't motivate the sphere or cone formulas because they require calculus. However, we can motivate the cylinder. Volume = area of base times height, $\pi r^2(h)$. Surface area: top and bottom are circles. Now the side. Imagine a soup can. If you take a knife, make a lengthwise cut, and take the label off—I still do this; how about you?—you get . . . a rectangle!!!! Area is base times height. The height is the height of the can if you neglect the rim, and the base is . . . the circumference of a circle, $2\pi r$. SA = $2\pi r(h)$. I once had a neighbor who wanted to know this formula. Of course, being a teacher I had to tell him how to get it. Of course, he couldn't have cared less, but he had to listen. Maybe that's why he moved away. Seriously, I don't even remember who it was!!

OTHER GEOMETRIC FACTS

Words about the circle

O—*Center* of circle.

OA—*Radius* (plural *radii*).

COD—*Diameter:* d = 2r.

EF—*Chord:* A line segment from one side of the circle to the other. The diameter is the longest chord.

UV—*Secant:* A line through a circle hitting two points of a circle.

RS—*Tangent:* A line hitting the circle at one and only one point.

Angle OPS—*Right angle:* The radius joined to the point of tangency, P, always forms a right angle. ∡OPR is also 90°.

D͡AP—*Arc:* A minor arc because it is less than half a circle.

D͡EP—*Major arc* because it is more than half a circle.

C͡PD—*Semicircle.*

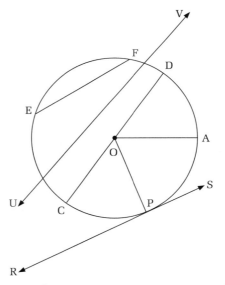

Arcs are measured in two ways. One way is degrees. There are 360° in one circle.

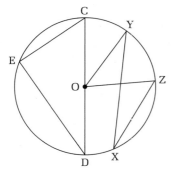

Angle YOZ is a *central angle.* It is equal in degrees to its intercepted arc Y͡Z. If ∡YOZ = 42°, so does arc Y͡Z.

Angle YXZ is an inscribed angle because its vertex hits the circle. Inscribed angles equal half the intercepted arc. Because Y͡Z = 42°, ∡YXZ = 21°.

$\overset{\frown}{CED}$ is a semicircle. ⊾CED is 90° because it is half the intercepted arc of a semicircle (180°). An inscribed angle in a semicircle is *always* a right angle.

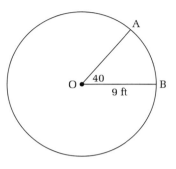

Arc lengths are lengths and are measured in inches, meters, and so on. Given sector AOB (the pie shape), find the arc length $\overset{\frown}{AB}$. The letter that is used is s.

Arc AB is 40/360 of the circle. In general

$$s = \frac{\text{central angle}}{360} (2\pi r) = \frac{40}{360} 2\pi(9) = 2\pi \text{ feet}$$

The perimeter $p = s + 2r = 2\pi + 2(9) = 2\pi + 18$

Area of the sector $= \dfrac{\text{angle}}{360} (\pi r^2) = \dfrac{40}{360} \pi(9)^2 = 9\pi$ square feet

If one of the following is true, all are true:

1. $\overline{OB}$ perpendicular to $\overset{\frown}{AC}$.
2. $\overline{OB}$ bisects $\overline{AC}$.
3. $\overline{OB}$ bisects arc $\overset{\frown}{ABC}$.

If line l is the perpendicular bisector of the chord $\overline{XY}$, line l goes through the center of the circle.

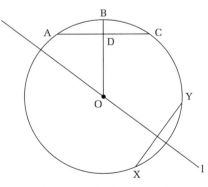

LINEAR INEQUALITIES

We would like to do some inequalities. Books use different notations to indicate intervals. Let me list some for you.

The Words	Notations	Pictures
x is greater than 4.	$x > 4$ $(4,\infty)$ Parenthesis on 4 means 4 is *not* part of the answer	Open dot, not including 4.
x is less than or equal to –6.	$x \leq -6$ $(-\infty,-6]$ Bracket near –6 means –6 IS part of the answer	Closed dot includes –6.
x is greater than 2 and less than 9. x is between 2 and 9. x is more than 2 *and* less than 9.	$2 < x < 9$ $(2,9)$	
x is between –4 and 6 inclusive. –4 is greater than or equal to x and less than or equal to 6.	$-4 \leq x \leq 6$ $[-4,6]$	
x is between 3 and 7 including 7.	$3 < x \leq 7$ $(3,7]$	
x is greater than or equal to 8 *or* less than 3.	$x \geq 8$ or $x < 3$ $[8,\infty)$ or $(-\infty,3)$	

Linear inequalities are very similar to linear equations. Let's try some. They are not too bad.

DEFINITION

$a < b$, "a is less than b": a is to the left of b on the number line.

$a > b$, "a is greater than b": a is to the right of b on the number line.

$c < d$ is the same as $d > c$.

DEFINITION

a ≤ b means a < b or a = b.

Make sure you understand this:

4 ≤ 9?   Yes, because 4 < 9.

6 = 6?   Yes, because 6 = 6.

−2 < −6?   No, −2 > −6 (look at number line). *Also,*
a ≥ b means a > b or a = b.

There are certain laws for inequalities.

1. *Trichotomy* law (*tri* means three and *chot* means cut): Exactly one is true: a < b, a = b, or a > b. If you compare two numbers, either they are equal, or one is larger than another. I know this is obvious, but it must be said. Another way to say this is a > 0, a = 0, or a < 0.

2. *Transitivity:* If a < b and b < c, then a < c. If a is less than b and b is less than c, then a is less than c. (The same is true for ≤, >, and ≥, but I won't say the same law 4 times!!!)

3. If a < b, then the following are true (similarly for ≤, >, and ≥):

   a. a + c < b + c   If you add the same to both sides, the order of the inequality stays the same.

   b. a − c < b − c   If you subtract, the order stays the same.

   c. if c > 0, ac < bc, and a/c < b/c   If you multiply or divide by a positive, the order stays the same.

   d. if c < 0, ac > bc and a/c > b/c   If you multiply or divide by a negative number, the order reverses.

Now let's use numbers. $4 < 10$. Then

   a. $4 + 3 < 10 + 3$

   b. $4 - 3 < 10 - 3$

   c. $4(2) < 10(2)$ and $\dfrac{4}{2} < \dfrac{10}{2}$ but

   d. $4(-2) > 10(-2)$

      $-8 > -20$ (Order switches.)

EXAMPLE 1—

Solve for x and graph:

$3x - 6 \le 7x - 30$

We solve this exactly the same way we do an equality, except if you multiply or divide by a negative the order changes.

$$3x - 6 \le\ 7x - 30$$
$$-7x\ \ \ \ \ = -7x$$
$$-4x - 6 \le\ \ \ \ \ \ -30$$
$$+6 =\ \ \ \ \ \ +6$$
$$-4x \le\ \ \ \ \ \ -24$$
$$\dfrac{-4x}{-4} \ge \dfrac{-24}{-4} \qquad \text{Dividing by } -4 \text{ switches the order.}$$
$$x \ge 6$$

The graph is a one-dimensional graph. The dot is the equal sign. If we have $\le$ or $\ge$, then equal is part of the answer and the dot is *solid*. If we have $<$ or $>$, the equal is *not* part of the answer and the dot is *open*. So the graph is

      6

EXAMPLE 2—

Solve for x and graph:

$2x + 5 \le 4x - 3 < x + 27$

We have 2 inequalities to solve:

$2x + 5 \le 4x - 3$ annnd $4x - 3 < x + 27$

$\qquad 5 \le 2x - 3 \qquad\qquad\qquad 3x - 3 < \qquad 27$

$\qquad 8 \le 2x$ annnd $3x \quad < \qquad 30$

$\qquad 4 \le x \qquad\qquad\qquad\qquad x \quad < \qquad 10$

The only way x can be greater than or equal to 4 and at the same time less than 10 is if x is between 4 and 10, including 4. The graph is

EXAMPLE 3—

Solve for x and graph:

$$3 \le \frac{5(-3x - 5)}{-2} \le 5$$

Because the x is in the middle only, we can do both inequalities at once.

$3(-2) \ge 5(-3x - 5) \ge 5(-2)$ **Multiplying by a negative reverses the order.**

$\qquad -6 \ge -15x - 25 \ge -10$
$\quad +25 \qquad\qquad +25 \quad +25$

$\qquad 19 \ge -15x \ge 15$

$$\frac{19}{-15} \le \frac{-15x}{-15} \le \frac{15}{-15}$$

Dividing by a negative again reverses the order!

$$-\frac{19}{15} \le x \le -1$$

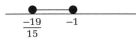

And the answer is all real numbers between −19/15 and −1 inclusive (including both ends). Remember, all of the graphs include all integers, rationals and irrationals, everything, between the two ends.

VARIATIONS OF VARIATION

There are three kinds of variation. Let's look at all three.

Direct variation: If y varies directly as x, we write y equals a constant times x. We write y = kx. k is used for a constant because, as we know, mathematicians can't spell.

EXAMPLE 1—

y varies directly as x. If y = 10 when x is 2, find y if x = 7.

y = kx. So 10 = k(2). We get the value of k, 5.

y = 5x. Now if x = 7, y = 2(7) = 14. Hmmmm. Not too bad. Let's do another type.

Inverse variation: If y varies inversely as x, we write y = k/x.

EXAMPLE 2—

y varies inversely as x. If y = 8 when x = 9, find y if x = 2.

If y = 8 and x = 9,
$$8 = \frac{k}{9}$$

So k = 72.
$$y = \frac{72}{x}$$

If x = 2, then
$$y = \frac{72}{2} = 36$$

Joint variation occurs when you have two or more of any kinds of variation.

EXAMPLE 3

y varies directly as the square of x and inversely as z. If y = 10 when x is 4 and z is 8, find y if x is 6 and z = 2.

$$y = \frac{kx^2}{z} \qquad 10 = \frac{k4^2}{8} \qquad 16k = 80$$

So $k = \frac{80}{16} = 5$

Now $y = \frac{5x^2}{z}$

So $y = \frac{5(6)^2}{2} = 90$

Some topics in this book are not so easy. This is *not* one of them. Enjoy it!

FUNCTIONS

One of the most neglected topics in high school is the study of functions. So let's get started at the beginning.

Function: Given a set D. To each element in D, we assign one and only one element.

EXAMPLE 1

Does the picture here represent a function? The answer is yes. 1 goes into a, 2 goes into 3, 3 goes into 3, and 4 goes into *pig*. Each element in D is assigned one and only one element.

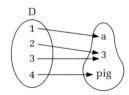

The next example will show what is not a function. But let us talk a little more about this example. The set D is called the *domain.* We usually think about x values when we think about the domain. This is not necessarily true, but it is true in nearly all high school and college courses, so we will assume it.

There is a second set that arises. It is not part of the definition. However, it is always there. It is called the *range.* Notice that the domain and the range can contain the same thing (the number 3) or vastly different things (3 and *pig*). However, in math, we deal mostly with numbers and letters. The rule (the arrows) is called the *map* or *mapping.* 1 is mapped into a; 2 is mapped into 3; 3 is mapped into 3; and 4 is mapped into *pig*.

Functional Notation

The rule is usually given in a different form: f(1) = a (read "f of 1 equals a"); f(2) = 3; f(3) = 3; and f(4) = *pig*.

NOTE 1
We cannot always draw pictures of functions, and we will give more realistic examples after we give an example of something that is not a function.

NOTE 2
The set of elements in the domain is sometimes called the set of *preimages* (what the elements looked like before the set is mapped). After it is mapped (and Hollywood makes them over), the elements in the ranges are its *images.*

EXAMPLE 2—

The picture here does not represent a function, since 1 is assigned two values, a and d.

EXAMPLE 3—

Let $f(x) = x^2 + 4x + 7$. $D = \{1, -3, 10\}$.

$f(1) = (1)^2 + 4(1) + 7 = 12$

$f(-3) = (-3)^2 + 4(-3) + 7 = 4$

$f(10) = (10)^2 + 4(10) + 7 = 147$

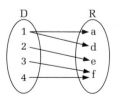

NOTE

When we think of the range, we will think of the y values, although again this is not necessarily true.

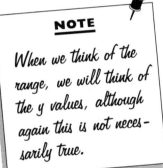

The range would be {4,12,147}. If we graphed these points, we would graph (1,12), (−3,4), and (10,147).

NOTE
Instead of graphing points (x,y), we are graphing points (x,f(x)). For our purposes, the notation is different, but the meanings are the same.

EXAMPLE 4—

Let $g(x) = x^2 − 5x − 9$. D = {4,0,−3,a^4, x + h}. Find the elements in the range.

This is a pretty crazy example, but there are reasons to do it.

$$g(4) = (4)^2 − 5(4) − 9 = −13 \qquad g(0) = 0^2 − 5(0) − 9 = −9$$

$$g(−3) = (−3)^2 − 5(−3) − 9 = 15 \qquad g(a^4) = (a^4)^2 − 5a^4 − 9$$

$$= a^8 − 5a^4 − 9 \qquad g(x + h) = (x + h)^2 − 5(x + h) − 9$$

Wherever there is an x, you replace it by x + h!

$$= x^2 + 2xh + h^2 − 5x − 5h − 9$$

The range is {−13, −9, 15, $a^8 − 5a^4 − 9$, $x^2 + 2xh + h^2 − 5x − 5h − 9$}.

EXAMPLE 5—

$f(x) = x/(x + 5)$.

Find

$$\frac{f(x + h) − f(x)}{h}$$

$$\frac{f(x + h) − f(x)}{h} = \frac{\dfrac{x + h}{x + h + 5} − \dfrac{x}{x + 5}}{h}$$

Add the fractions. Two tricks: a/b − c/d = (ad − bc)/bd; (e/f)/h = e/fh.

$$= \frac{(x + h)(x + 5) − x(x + h + 5)}{(x + h + 5)(x + 5)h}$$

Multiply out the top; never multiply out the bottom.

Cancel the h's.

$$= \frac{5h}{(x + h + 5)(x + 5)h} = \frac{5}{(x + h + 5)(x + 5)}$$

Why do we want to study something crazy like this? Shhh. It's a secret. This is one of the first things we look at in calculus! Shhh. Don't tell anyone.

Let's look at a couple of functions.

EXAMPLE 6—

$f(x) = 3x + 1$

This should look familiar. It looks like $y = 3x + 1$. If you graph it, you will find that it is. It is an oblique (slanted) straight line. Any such line has domain and range, all real numbers. Let's graph it.

x	f(x)	
−1	−2	(−1,−2)
0	1	(0,1)
2	7	(2,7)

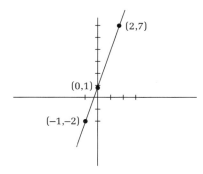

Let's do a little on the parabola. We'll look at the parabola $f(x) = ax^2 + bx + c$. The shape of the parabola is determined by the coefficient of x.

If a > 0, its shape is

V

If a < 0, its shape is

V

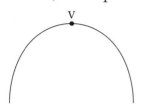

The low point or high point, indicating the letter V, is the *vertex*.

The x coordinate of the vertex is found by setting x equal to $-b/(2a)$. The y value is gotten by putting the x value into the equation for the parabola. The line through the vertex, the axis of symmetry, is given by $x = -b/(2a)$. The intercepts plus the vertex usually are enough for a fairly good picture. We will do some now.

EXAMPLE 7—

$f(x) = 2x^2 - 7x + 3$

Vertex $x = \dfrac{-b}{(2a)} = \dfrac{-(-7)}{2(2)} = \dfrac{7}{4}$

$y = 2\left(\dfrac{7}{4}\right)^2 - 7\left(\dfrac{7}{4}\right) + 3 = \dfrac{-25}{8}$ $\qquad \left(\dfrac{7}{4}, \dfrac{-25}{8}\right)$

Axis of symmetry: $x = \dfrac{-b}{(2a)}$ $\qquad x = \dfrac{7}{4}$

y intercept: $x = 0$, $y = 3$ $\qquad (0,3)$

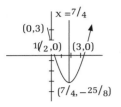

x intercepts: y = 0

$2x^2 - 7x + 3 = (2x - 1)(x - 3) = 0$

x = 1/2, 3 and the intercepts are (1/2,0) and (3,0). The graph opens upward.

EXAMPLE 8—

$f(x) = x^2 + 6x$

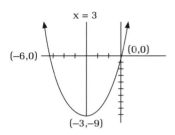

Vertex $x = \dfrac{-b}{(2a)} = \dfrac{-6}{2}$ (1) = -3

$y = (-3)^2 + 6(-3) = -9$ $(-3,-9)$

x intercepts: y = 0

$x^2 + 6x = x(x + 6) = 0$

x = 0, −6—(0,0); also, the y intercept is (−6,0). Picture is up. Axis of symmetry: x = −3.

EXAMPLE 9—

$f(x) = 9 - x^2$

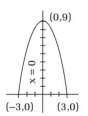

Vertex $x = \dfrac{-b}{(2a)} = \dfrac{0}{2}$ (−1) = 0

y = 9   (0,−9) also y intercept

x intercepts: y = 0

$0 = 9 - x^2 = (3 - x)(3 + x)$

x = 3, −3; intercepts are (3,0) and (−3,0). Picture is down. Axis of symmetry: x = 0.

EXAMPLE 10—

$f(x) = x^2 - 2x + 5$

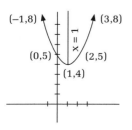

Vertex $x = -\dfrac{-2}{2(1)} = 1$

$y = 1^2 - 2(1) + 5$ (1,4)

The y intercept is (0,5). $y = x^2 - 2x + 5 = 0$. The quadratic formula gives imaginary roots, so no x intercepts.

To get more points, make a chart. Take two or three x values (integers) just below the vertex and two or three just above.

x	y
−1	8
0	5
1	4
2	5
3	8

Axis of symmetry: x = 1. Parabola is up.

More about parabolas? You can find it in *Precalc with Trig for the Clueless!*

Let's look at composite functions.

Composite Functions

Suppose we have a function, map f, whose domain is D and whose range is R_1. Suppose also we have another function, map g, domain R_1 and range R_2. Is there a map that goes from the original domain D to the last range R_2? The answer is, of course, yes; otherwise why would I waste your time writing this paragraph?

DEFINITION
Given a function, map f, domain D, range R_1. Given a second function, map g, domain R_1, range R_2. Define the composite map g ∘ f, domain D, range R_2.
(g ∘ f)(x)—read "g circle f of x"—is g[f(x)], read "g of f of x." The picture might look as shown here:

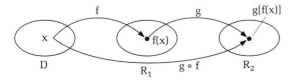

EXAMPLE 11

Suppose $f(x) = x^2 + 4$ and $g(x) = 2x + 5$.

a. $g(f(3))$. $f(3) = 3^2 + 4 = 13$. $g(13) = 2(13) + 5 = 31$.
The picture is as shown here:

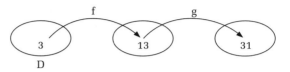

b. $f[g(3)]$. First, let's note that the picture would be totally different. Second, the "inside" map always is done first. $g(3) = 2(3) + 5 = 11$. $f(11) = 11^2 + 4 = 125$. Rarely are the two composites the same.

c. $g(f(x))$. $g(x) = 2x + 5$. This means multiply the point by 2 and add 5. $g(f(x))$ means the point is no longer x but $f(x)$. The rule is multiply $f(x)$ by 2 and add 5. So $g(f(x)) = 2(f(x)) + 5 = 2(x^2 + 4) + 5 = 2x^2 + 13$.

NOTE

$g(f(3)) = 2(3)^2 + 13 = 31$, which agrees with item a above.

d. $f(g(x))$. $f(x) = x^2 + 4$. So $f(g(x)) = (g(x))^2 + 4 = (2x + 5)^2 + 4 = 4x^2 + 20x + 29$. $f(g(3)) = 4(3)^2 + 20(3) + 29 = 125$, which agrees with item b, as it must.

Inverse Functions

There is a function that takes the domain into the range. Is there a function that takes the range back into the domain? The answer is yes under certain circumstances. That circumstance is when the domain and range are in 1:1 correspondence. This means that every element in the domain can be paired off with one and only one element in the range. In the case of a finite set, 1:1 correspondence means the same number of elements.

On a graph you can see whether you have an inverse by first using the vertical line test (each line hits the graph only once) to see whether the curve is a function. Then use the horizontal line test (each line also hits the curve only once) to determine whether there is an inverse function. The most common kinds of functions that have inverses are ones that always increase or always decrease.

The definition of an inverse function is very long, but it is not very difficult after you read it and the examples that follow:

Given a function, map f, domain D, range R, D, and R are in 1:1 correspondence. Define f^{-1}, read "f inverse," domain R, range D. If originally a was an element of D and b was an element of R, define $f^{-1}(b) = a$ if originally $f(a) = b$.

This looks quite bad, but really it is not. Look at the picture here.

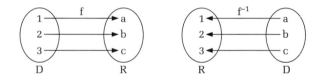

$f^{-1}(a) = 1$ because originally $f(1) = a$; $f^{-1}(b) = 2$ because $f(2) = b$; and $f^{-1}(c) = 3$ because $f(3) = c$.

A little better? Let's make it much better. First note the inverses we know: adding and subtracting, multiplying and dividing (except with 0), squaring and square roots, providing only one square root, cubing and cube roots, and so on.

EXAMPLE 12—

Given map $f(x) = 2x + 3$, domain {1,8,30}.

$f(1) = 5$, $f(8) = 19$, $f(30) = 63$. The range is {5,19,63}. D and R are in 1:1 correspondence. Let's find the inverse. The domain and range switch. The new domain is

{5,19,63}. Let's find the new map, f^{-1}. $f(x) = 2x + 3$ means multiply by 2 and add 3. Going in the opposite direction, adding 3 was last so subtracting 3 is first. First we multiply by 2; going backward, the last thing is dividing by 2.

$f(x) = 2x + 3.2x + 3 = f(x)$. $2x = f(x) - 3$. $x = [f(x) - 3]/2$. Notation change: $f(x)$ is now in the new domain. Change $f(x)$ to x! x is the new function; so x becomes $f^{-1}(x)$. So our new function is $f^{-1}(x) = (x - 3)/2$, D = {5,19,63}. Let's check it out!! $f^{-1}(5) = (5 - 3)/2 = 1$. Okay so far. $f^{-1}(19) = (19 - 3)/2 = 8$. $f^{-1}(63) = (63 - 3)/2 = 30$. Everything is okay.

Let's do some more examples.

EXAMPLE 13

The domain is all real numbers except x = 3, because the bottom of a fraction may never be equal to 0. However, it is not clear at all what the range is. Let's find the inverse function. Perhaps the original range and the new domain will become more obvious. In order to make this less messy, let f(x) = y.

$$f(x) = \frac{x + 2}{x - 3}$$

Multiply through by x − 3.

$$\frac{y}{1} = \frac{x + 2}{x - 3}$$

$$y(x - 3) = x + 2$$

$$yx - 3y = x + 2$$

$$yx - x = 3y + 2$$

$$x(y - 1) = 3y + 2$$

$$x = \frac{3y + 2}{y - 1}$$

At this point it is clear that the old range was all y values except for I. Now the notation changes.

Recall: Notation change: y becomes the x (in the new domain). The new function x is replaced by $f^{-1}(x)$, the inverse of the original.

$$f^{-1}(x) = \frac{3x + 2}{x - 1}$$

EXAMPLE 14—

Find the inverse map for

$$f(x) = \sqrt{3x + 1}$$

with the domain $x \geq 1$ and the range $y \geq 2$. Let $f(x) = y$.

$$y = \sqrt{3x + 1}$$

$$\sqrt{3x + 1} = y$$

$$3x + 1 = y^2$$

$$3x = y^2 - 1$$

$$x = \frac{(y^2 - 1)}{3}$$

So $f^{-1}(x) = \dfrac{(x^2 - 1)}{3}$, $x \geq 2$

NOTE:

Sometimes, in order to find the range, it is a good idea to solve for x!!!!

TRANSLATIONS, FLIPS, STRETCHES, AND CONTRACTIONS

This topic has become increasingly popular over the years. It is not to my liking, since it is presented much better later on.

Suppose F(x) looks like this:

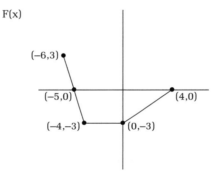

F(x) + 2 means 2 up:

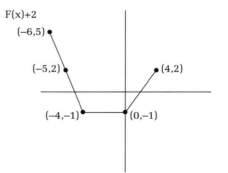

F(x) − 3 is 3 down:

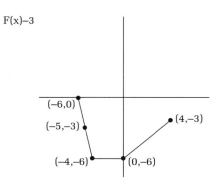

F(x − 4) is 4 to the right:

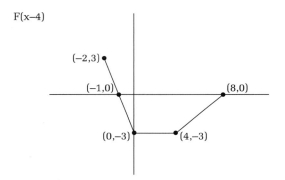

F(x + 1) is 1 to the left:

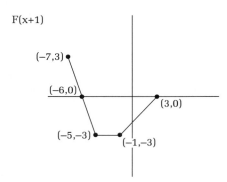

2F(x) is twice as high:

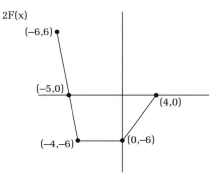

(1/3)F(x) is ⅓ the height:

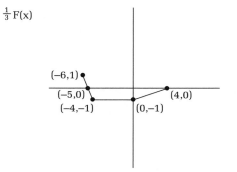

−F(x) is upside down:

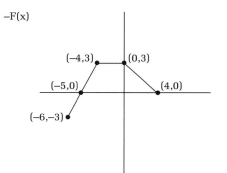

F(–x) is a reflection in the y axis:

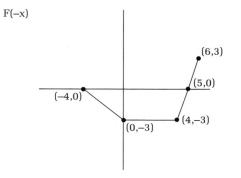

F(3x) is ⅓ the width:

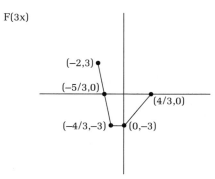

F(½x) is twice the width:

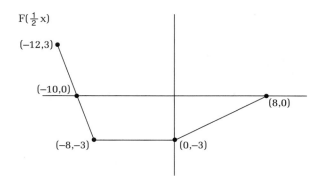

$-2F(x - 4) + 7$:

a. 4 to the right

F(x–4)

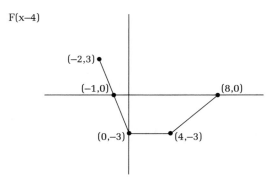

b. Upside down and twice the height:

–2F(x–4)

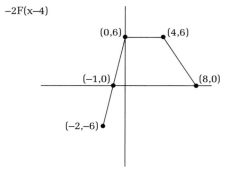

c. 7 up:

–2F(x–4)+7

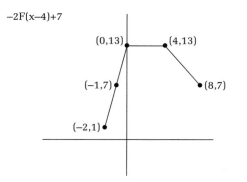

PREIMAGES AND IMAGES REVISITED

Preimages and images are starting to be included in basic algebra and on the SAT in a new way. Since I'm sure my publisher won't allow a new edition in a year or two (too expensive), I'll include the information here.

EXAMPLE I—

Suppose we have a function f(x). The *preimage* is always the point (5,−11). What is the *image,* given the following functions?

The interpretation will be given together with the answer.

   a. f(x) + 2: up 2, in the y direction: (5,−9)

   b. f(x + 2): 2 to the left, in the x direction: (3,−11)

   c. f(x) − 6: down 6, in the y direction: (5,−17)

   d. f(x − 6): 6 to the right, in the x direction: (11, −11)

   e. −f(x): upside down, the sign of the y number changes: (5,11)

   f. f(−x): reflection in the x axis, the sign of x changes: (−5,−11)

   g. 3f(x): 3 times the height, the y number: (5,−33)

   h. (−½)f(x): half the height, the y number and upside down: (5,11/2)

   i. f(7x): 7 times the compression of the x number: (5/7,−11)

   j. f(⅓x): stretch, 3 times the x number: (15,−11)

   k. 5f(x − 2) + 6: We start with the point (5,−11). First we go 2 to the right: (7,−11). Next we go 5 times the height: (7,−55). Finally we go up 6: (7,−49).

EXAMPLE 2—

Suppose we have a function f(x) and the *image* is (5, −11). Find the *preimage,* given the following functions:

a. f(x) + 2: Since this means add 2 to the height, the preimage must be 2 less in the y direction: (5,−13)

b. f(x + 2): This is 2 to the left. The preimage must be 2 to the right: (7,−11)

c. f(x) − 6: 6 down; the preimage must be 6 up (almost a good name for a soda) (7,−5)

d. f(x − 6): 6 right; the preimage is 6 left: (−1,−11)

e. −f(x): upside down; interestingly, it is the same as Example 1: (5,11)

f. f(−x): sign of the x number changes; again, the same as in Example 1: (−5,−11)

g. 3f(x): height 3 times, preimage must be ⅓: (−5, −11/3)

h. (−½)f(x): half the height, upside down: y number changes sign and the preimage must be twice as high and upside down: (5,22)

i. f(7x): x values compression of ⅐: preimage must be 7 times in distance: (35,−11)

j. f(⅓x): x stretches three times; preimage must be ⅓; (5/3, −11)

k. 5f(x − 2) + 6: The directions say: go 2 to the left, 5 times as high, up 6. The preimage is the reverse in the reverse order. We start with the point (5, −11). The last thing we do is go 6 up. So the first thing we do in reverse is go down 6: (5,−17). The second part is 5 times as high. The preimage y value is ⅕ as high. (5,−17/5). The first thing we did is 2 to the right. So the last thing in reverse is to go 2 to the left: $(3, -17/5)$.

If you want to know more about functions and other algebraic topics, get *Calc for the Clueless* and *Precalc with Trig for the Clueless*. The SAT? *SAT® Math for the Clueless!*

As for now, thththat's all folks!

CHAPTER 11

MISCELLANEOUS MISCELLANY

This chapter has been added because most schools include these topics in basic algebra. For totally different reasons, I wish that none of these sections would be in your basic algebra class. The first section is absolute value. It seems very easy but is not. It is also not needed until later. Unfortunately, most schools cover it much too early.

The second section involves counting, probability, and statistics. This is a wonderful section to study, after all the basics are mastered. However, these topics are used to replace more necessary work.

The third section is matrices. This is a complete waste of time at this level, since the first time you actually need matrices is about one year after you finish the three-semester college calculus sequence.

ABSOLUTE VALUE

There are two definitions of absolute value. Both are needed.

DEFINITION I

$|x| = \sqrt{x^2}$. $|8| = \sqrt{8^2} = \sqrt{64} = 8$. $|0| = \sqrt{0^2} = \sqrt{0} = 0$.
$|-4| = \sqrt{(-4)^2} = \sqrt{16} = 4$.

Although this is the easier definition, the other is more useful.

DEFINITION 2

$$|x| = \begin{array}{l} x \text{ if } x > 0 \\ -x \text{ if } x < 0 \\ 0 \text{ if } x = 0 \end{array}$$

$|8| = 8$, since $8 > 0$. $|-4| = -(-4) = 4$, since $-4 < 0$.

Read this definition very, very carefully.

NOTE:

$|x|$ is never negative.

EXAMPLE IA—

Solve for x:

$|3x - 1| = 8$

$|u| = 8$ means $u = \pm 8$. We have $3x - 1 = 8$ or $3x - 1 = -8$. So $x = 3$ or $-7/3$.

For fun, let's do the problem with the other definition.

EXAMPLE IB—

Solve for x:

$|3x - 1| = 8$

$|3x - 1| = \sqrt{(3x - 1)^2} = 8$. $(3x - 1)^2 = 9x^2 - 6x + 1 = 64$ or $9x^2 - 6x - 63 = 0$.

Factoring, we get $9x^2 - 6x - 63 = 3(3x^2 - 2x - 21) = 3(3x + 7)(x - 3) = 0$.

$3x + 7 = 0$ or $x = -7/3$; $x - 3 = 0$ or $x = 3$.

This way is much longer, but sometimes it is the only

way. However, this method is not generally part of a basic course.

EXAMPLE 2—

Solve for x:

$|9x - 5| = 0$

$|u| = 0$ only when $u = 0$. $9x - 5 = 0$. So $x = 5/9$.

EXAMPLE 3—

Solve for x:

$|4x + 5| = -2$

There is no solution, since the absolute value is never negative.

EXAMPLE 4—

Solve for x and graph:

$|x - 10| \leq 3$

If $|u| \leq 3$ and u were an integer, the u could equal -3, $-2, -1, 0, 1, 2, 3$.

In other words, $-3 \leq x \leq 3$. Solving, we get:

$-3 \leq x - 10 \leq 3$

$+10 + 10 + 10$ (add 10 to all three parts)

$7 \leq x \leq 13$

Its graph looks like this:

The solid dots mean that both 7 and 13 are part of the answer (the equal signs).

EXAMPLE 5—

Solve for x and graph:

$|2x - 7| < 5$

$-5 < 2x - 7 < 5$

$\phantom{-5 <} 7 \phantom{< 2x -} 7 7$

$\phantom{-5 <} 2 < \phantom{<} 2x < 12$

$\phantom{-5 <} 1 < \phantom{< 2} x < 6$

Its graph looks like this:

The open dots mean that 1 and 6 are *not* part of the answer (no equal sign). However, 5.9999999 and 1.00000001 are part of the answer.

EXAMPLE 6—

Solve for x and graph:

$|x - 7| \geq 3$

If $|u| \geq 3$ and u is an integer, $u = 3, 4, 5, 6, \ldots$ and $-3, -4, -5, -6, \ldots$.

In other words, $|u| \geq 3$ means $u \geq 3$ or $u \leq -3$.

Solving our problem, $x - 7 \geq 3$ or $x - 7 \leq -3$. So $x \geq 10$ or $x \leq -4$. Its graph looks like this:

EXAMPLE 7—

Solve for x and graph:

$|5 - 2x| > 9$

$ 5 - 2x > 9 \quad$ or $\quad 5 - 2x < -9$

$ -2x > 4 \quad$ or $\quad -2x < -14$

So
$$x < -2 \quad \text{or} \quad x > 7$$

EXAMPLE 8—

Solve for x:

$$|3x - 5| < -4$$

There is no answer, since the absolute value is never negative. Teachers have been known to put this question on exams to test understanding.

EXAMPLE 9—

Solve for x:

$$|5x + 4| > -2$$

The answer is all real numbers, since the absolute value is never negative and is always bigger than a negative number.

Let's look at counting, probability, and statistics.

COUNTING, PROBABILITY, AND STATISTICS

Statistics

Let's take a look at statistics first, since most of you have done this a little.

There are three words that mean average: *mean, median,* and *mode.*

DEFINITION

Mean: Add up all the numbers and divide by that number. This is the way most of your grades are determined.

DEFINITION
Median: The middle number.

DEFINITION
Mode: The most common number.

EXAMPLE 1—

Determine the mean, median, and mode for 99, 87, 87, 93, 98, 97, 96.

First, arrange the grades from lowest to highest: 87, 87, 93, 96, 97, 98, 99.

The *range* (values from the lowest to highest) is 87 to 99.

The median is the middle grade. Since there is an odd number (7 in this case), the median is the fourth grade: 96.

The mode is the most common grade: 87.

The mean is $(99 + 87 + 87 + 93 + 98 + 97 + 96)/7 = 93\%$.

EXAMPLE 2—

Determine the range, median, mean, and mode for 40, 40, 45, 48, 52, 52.

Range: 40 to 52. Mode: 40 and 52 (*bimodal*—two modes). There can be any number of modes.

Mean: $(40 + 40 + 45 + 48 + 52 + 52)/6 = 46\frac{1}{6}$

Median: Since there is an even number, the median is the average (mean) of the middle two: $(45 + 48)/2 = 46.5$

If you have to take a survey, which "average" (median, mean, or mode) is best statistically?

A STORY

It may surprise you to learn that the answer is the median and not the mean. The story is fictional, but there is actual reality to it. If I told you the average person in a country made over $1 million a year, what would you say about the people? You might say everyone is rich! That's not really true. Suppose we have a small island republic with 10 people on it. The incomes of 9 of the people are $70, $75, $80, $85, $85, $90, $90, $90, $95. The tenth person, the owner of the oil well, makes $100,000,000 a year. Add them up and divide by 10 and the average income is more than $10,000,000 a year! Is everyone rich? No! The mean is distorted by extremes. The median is $87.50, a much better determination of how much the "average" person makes. This story demonstrates why people say statistics can be used to prove anything. You have to know much about the statistics, including the group you are taking statistics about, to understand what the statistics say.

Counting

This seems to be a very silly topic, since everyone can count. Sometimes counting is very easy. However, sometimes it is not.

Basic *Law of Counting:* If you can first do something in m ways, the next thing in n ways, the next thing in p ways, and so on, the total number of ways you can do something in that order is m $\times$ n $\times$ p $\times$ . . . .

EXAMPLE 1

Sandy has a choice of five sandwiches, four desserts, and three drinks. How many different meals are there if a meal consists of one of each?

The number of different meals is $5 \times 4 \times 3 = 60$.

Okay, okay. If this whole topic were that easy, it wouldn't be here. Let's do more.

EXAMPLE 2—

Suppose we have the letters a, b, c, d, e. A three-letter "word" is any three letters together. So bba is a word.

  i. How many three-letter words are there? Five choices for the first letter, five for the second (you can repeat), five for the third letter: $5 \times 5 \times 5 = 125$ words.

 ii. How many three-letter permutations (no repetition)? $5 \times 4 \times 3 = 120$ words.

iii. How many three-letter word permutations with the first letter a vowel and the second letter a consonant? There are two vowels, a and e. Two choices for the first letter. There are three consonants, c, d, and e. So there are three choices for the second letter. Since this is a permutation, we have used up one vowel and one consonant. There are only three choices for the third letter. The total is $2 \times 3 \times 3 = 18$. If this were not a permutation, the answer would be $2 \times 3 \times 5 = 60$.

 iv. How many words with all three letters the same? There are five choices for the first letter, but only one choice for the second and the third, since they must all be the same. The total number is $5 \times 1 \times 1 = 5$. The words are aaa, bbb, ccc, ddd, and eee.

This is not quite as easy as it looks.

EXAMPLE 3—

How many ways can five people sit in a line?

This question is actually a permutation, since one person can't sit in two seats.

There are $5 \times 4 \times 3 \times 2 \times 1 = 120$ choices.

NOTE 1

$5 \times 4 \times 3 \times 2 \times 1$ can be written as 5! (read: "5 factorial").

NOTE 2

If there were n people, then the answer would be n! = $n(n-1)(n-2) \ldots 3 \times 2 \times 1$.

NOTE 3

0! = 1. It is illogical, but it makes all of the formulas true.

NOTE 4

If there are five people in a circle, there are five repeats for each position. So the answer is 5!/5 = 4!. You might draw the picture.

NOTE 5

In a circle, n people would have (n − 1)! different ways to sit.

Probability

We hear about probability all the time. Let's see what it is.

The probability of something happening is the number of happening events divided by the number of all the events.

EXAMPLE 1—

The weatherperson says there's a 30 percent chance of rain. What does that mean? It means that out of 100 identical days, 30 of those days it will rain.

EXAMPLE 2—

Given the 26-letter English alphabet. The vowels are a, e, i, o, and u. The rest are consonants.

a. Probability of a vowel: 5/26

b. Probability of a consonant: 21/26

c. Probability of picking an English letter: 26/26 = 1. The probability of a sure thing is 1.

d. Probability of picking the letter π: 0/26 = 0. The probability of something not happening is 0.

NOTE

Probability of a vowel = 1 − probability of not a vowel (consonant).

In general, the probability of happening + probability of not happening = 1.

FOR FUN

Suppose you are in a class or a group of 30 people. What are the chances of 2 people having the same birthday, month and day only, not the year? Would you guess a very bad chance, slightly bad chance, 50-50, slightly good chance, or very good chance? As strange as it seems, there is a very, very good chance that 2 people have the same birthday. Let me explain.

The way to do this is to calculate the probability of 2 people not having the same birthday and then take 1 − that to figure the possibility of 2 people having the same birthday.

If we look at the first person, the first person can't have the same birthday as anyone else (there isn't anyone else). The probability is 1 that the person won't match another. The second person can match only the first. So the probability of not matching is 364/365. The third person can match the first 2. So the probability of not matching would be $1 \times 364/365 \times 363/365$. If we continue, the probability of 2 people matching would be $1 - 364/365 \times 363/365 \times 362/365x. \ldots$

For a group of only 23 (!!!!!!!), the chances are 50-50 that 2 people will have the same birthday. In a group of 30, you would really expect 2 people to match. At 40, it is almost a sure thing and a probability to have two sets of matches or a triple match. Try it in your class or group.

The last problem shows you what makes probability easy and what makes it not so easy. The math in the problem is arithmetic—in this case, messy arithmetic—but not even algebra. The more difficult part is finding out what the problem is. For example, if I had asked what was the probability of two people having their birthday on Thanksgiving, this would be a different problem, since a specific day is not the same probability as any two birthdays matching. That is why some people like this topic a lot and some don't like it a lot.

MATRICES

Matrices (singular is *matrix*) are arrays of numbers or letters.

An m × n (read: "m by n") matrix has m rows and n columns.

EXAMPLE 1

$$\begin{bmatrix} a & b & c \\ d & e & f \end{bmatrix}$$

is a 2×3 matrix: two rows

$[a\ b\ c]$ and $[d\ e\ f]$,

and three columns

$$\begin{bmatrix} a \\ d \end{bmatrix}, \begin{bmatrix} b \\ f \end{bmatrix}, \text{and } \begin{bmatrix} c \\ f \end{bmatrix}$$

a is the first-row, first-column entry.

b is the first-row, second-column entry.

c is the first-row, third-column entry.

d is the second-row, first-column entry.

e is the second-row, second-column entry.

f is the second-row, third-column entry.

There are three operations you need to know: addition, scalar multiplication, and multiplication.

Add: The rows and the columns must be the same; add corresponding entries.

EXAMPLE 2

$$\begin{bmatrix} a & b & c & d \\ e & f & g & h \end{bmatrix} + \begin{bmatrix} s & t & u & v \\ w & x & y & z \end{bmatrix} = \begin{bmatrix} a+s & b+t & c+u & d+v \\ e+w & f+x & g+y & h+z \end{bmatrix}$$

EXAMPLE 3

$$\begin{bmatrix} 1 & -5 & 0 \\ 6 & 4 & 3 \end{bmatrix} + \begin{bmatrix} -2 & 5 & -4 \\ 2 & 1 & 6 \end{bmatrix} = \begin{bmatrix} -1 & 0 & -4 \\ 8 & 5 & 9 \end{bmatrix}$$

Scalar multiply: Multiply each entry with the number (letter) on the outside.

EXAMPLE 4—

$$k\begin{bmatrix} 3 & 4 \\ a & -2 \end{bmatrix} = \begin{bmatrix} 3k & 4k \\ ak & -2k \end{bmatrix}$$

Multiply: To multiply, the columns of the left matrix must be the same as the rows of the right matrix.

EXAMPLE 5A—

If we have a 3×5 matrix multiplied by a 5×2 matrix, the result is a 3×2 matrix.

EXAMPLE 5B—

Let us reverse. If we have a 5×2 matrix multiplying a 3×2 matrix, we cannot even multiply them. We see that matrix multiplication is *not* commutative.

EXAMPLE 6—

$$\begin{bmatrix} a & b & c & d \\ e & f & g & h \end{bmatrix} \times \begin{bmatrix} o & p & q \\ r & s & t \\ u & v & w \\ x & y & z \end{bmatrix}$$

The first is 2×4; the second is 4×3. We can multiply, since the columns of the first are the same as the rows of the second. The result is a 2×3.

First-row, first-column entry: ao + br + cu + dx

First-row, second-column entry: ap + bs + cv + dy

First-row, third-column entry: aq + bt + cw + dz

Second-row, first-column entry: eo + fr + gu + hx

Second-row, second-column entry: ep + fs + gv + hy

Second-row, third-column entry: eq + ft + gw + hz

The matrix would be

$$\begin{bmatrix} ao + br + cu + dx & ap + bs + cv + dy & aq + bt + cw + dx \\ eo + fr + gu + hx & ep + fs + gv + hy & eq + ft + gw + hz \end{bmatrix}$$

You cannot multiply the matrices in the reverse order.

EXAMPLE 7—

Let $A = \begin{bmatrix} 0 & 1 \\ 2 & 3 \end{bmatrix}$ and $B = \begin{bmatrix} 4 & 5 \\ 6 & 7 \end{bmatrix}$

Find AB and BA.

$$AB = \begin{bmatrix} 0 & 1 \\ 2 & 3 \end{bmatrix}\begin{bmatrix} 4 & 5 \\ 6 & 7 \end{bmatrix} = \begin{bmatrix} 0(4) + 1(6) & 0(5) + 1(7) \\ 2(4) + 3(6) & 2(5) + 3(7) \end{bmatrix} = \begin{bmatrix} 6 & 7 \\ 26 & 31 \end{bmatrix}$$

$$BA = \begin{bmatrix} 4 & 5 \\ 6 & 7 \end{bmatrix}\begin{bmatrix} 0 & 1 \\ 2 & 3 \end{bmatrix} = \begin{bmatrix} 4(0) + 5(2) & 4(1) + 5(3) \\ 6(0) + 7(2) & 6(1) + 7(3) \end{bmatrix} = \begin{bmatrix} 10 & 19 \\ 14 & 27 \end{bmatrix}$$

Notice that these matrices are not commutative.

DEFINITION

Equal matrices: Two matrices are equal if they have the same number of rows and columns and corresponding entries are the same.

DEFINITION

Zero matrix: All entries are zero.

DEFINITION

Identity matrix: A square matrix (rows = columns); the main diagonals are all ones; everything else is zeros.

EXAMPLE 8—

Suppose $A = \begin{bmatrix} a & b & c \\ d & e & f \end{bmatrix}$ and $B = \begin{bmatrix} j & k & l \\ m & n & o \end{bmatrix}$

Since the rows and columns are the same, A = B if $a = j$, $b = k$, $c = l$, $d = m$, $e = n$, and $f = o$.

EXAMPLE 9—

$$3 \times 3 \text{ O matrix} = \begin{bmatrix} 0 & 0 & 0 \\ 0 & 0 & 0 \\ 0 & 0 & 0 \end{bmatrix} \quad 2 \times 2 \text{ O matrix is } \begin{bmatrix} 0 & 0 \\ 0 & 0 \end{bmatrix}$$

EXAMPLE 10—

3×3 identity matrix $I = \begin{bmatrix} 1 & 0 & 0 \\ 0 & 1 & 0 \\ 0 & 0 & 1 \end{bmatrix}$

2×2 identity matrix $I = \begin{bmatrix} 1 & 0 \\ 0 & 1 \end{bmatrix}$

DEFINITION

Inverse matrix A^{-1} is the matrix such that $AA^{-1} = A^{-1}A = I$. Only square matrices can have inverses.

I couldn't leave this section without a few applications of matrices, although they are ugly.

EXAMPLE 11—

Suppose $A = \begin{bmatrix} a & c \\ b & d \end{bmatrix}$

and $ad - bc \neq 0$, then $A^{-1} = \dfrac{1}{ad - bc} \begin{bmatrix} d & -c \\ -b & a \end{bmatrix}$

EXAMPLE 12—

Suppose $B = \begin{bmatrix} 4 & 5 \\ 2 & 3 \end{bmatrix}$ Find B^{-1}.

$4(3) - 5(2) = 2$ (not 0). So the matrix has an inverse.

$B^{-1} = \dfrac{1}{2} \begin{bmatrix} 3 & -5 \\ -2 & 4 \end{bmatrix} = \begin{bmatrix} 3/2 & -5/2 \\ -1 & 2 \end{bmatrix}.$

Theorems (proven laws):

Let $A = \begin{bmatrix} a & c \\ b & d \end{bmatrix}$

0 is the zero 2×2 matrix and I is the identity 2×2 matrix.

   i. $A + 0 = 0 + A = A$ for all matrices A.

  ii. $A \times 0 = 0 \times A = 0$ for all matrices A.

iii. $A \times I = I \times A = A$ for all matrices A.

iv. If $ad - bc \neq 0$, $A A^{-1} = A^{-1} A = I$ for all matrices A.

It will not be proved, but you might want to show that this is true for matrix B or even for matrix A.

EXAMPLE 13—

$BB^{-1} = I$ is true if $B = \begin{bmatrix} 4 & 5 \\ 2 & 3 \end{bmatrix}$

We already showed that $B^{-1} = \begin{bmatrix} 3/2 & -5/2 \\ -1 & 2 \end{bmatrix}$

$BB^{-1} = \begin{bmatrix} 4 & 5 \\ 2 & 3 \end{bmatrix} \begin{bmatrix} 3/2 & -5/2 \\ -1 & 2 \end{bmatrix}$

$= \begin{bmatrix} 4(3/2) + 5(-1) & 4(-5/2) + 5(2) \\ 2(3/2) + 3(-1) & 2(-5/2) + 3(2) \end{bmatrix} = \begin{bmatrix} 1 & 0 \\ 0 & 1 \end{bmatrix} = I$

You might want to show that $B^{-1}B = I$.

Finally, there is one use for matrices now. You can solve two equations in two unknowns.

EXAMPLE 14—

Solve for x and y:

$4x + 5y = 22$

$2x + 3y = 12$

We can write this in matrix form $BZ = T$, where

$B = \begin{bmatrix} 4 & 5 \\ 2 & 3 \end{bmatrix}$, $Z = \begin{bmatrix} x \\ y \end{bmatrix}$, and $T = \begin{bmatrix} 22 \\ 12 \end{bmatrix}$

(You might want to show $BZ = T$.)

But if we multiply both sides by B^{-1}, we get $B^{-1}BZ = B^{-1}T$.

But $B^{-1}B = I$ and $IZ = Z$. So $Z = B^{-1}T$.

We found B^{-1} before. It is $\begin{bmatrix} 3/2 & -5/2 \\ -1 & 2 \end{bmatrix}$

$Z = B^{-1}T$ means $\begin{bmatrix} x \\ y \end{bmatrix} = \begin{bmatrix} 3/2 & -5/2 \\ -1 & 2 \end{bmatrix}\begin{bmatrix} 22 \\ 12 \end{bmatrix}$

$$= \begin{bmatrix} 3/2(22) + -5/2(12) \\ (-1)(22) + 2(12) \end{bmatrix} = \begin{bmatrix} 3 \\ 2 \end{bmatrix}$$

Since $\begin{bmatrix} x \\ y \end{bmatrix} = \begin{bmatrix} 3 \\ 2 \end{bmatrix}$

$x = 3$ and $y = 2$

To check, $4(3) + 5(2) = 22$ and $2(3) + 3(2) = 12$. Whew! Too much work!!!!

NOTE

We could do everything for matrices that are 3×3, 4×4, etc., but even 3×3 is so long and so messy it won't be done here.

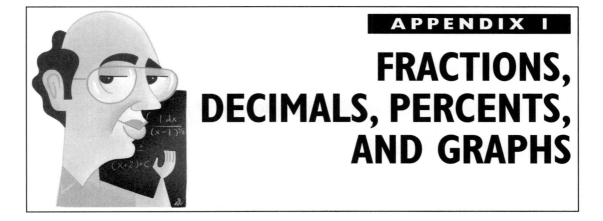

FRACTIONS, DECIMALS, PERCENTS, AND GRAPHS

If you read Chap. 1, you know how important it is to know your multiplication tables without using a calculator. It is also very important to know your fractions, decimals, and percents. If you don't know these without calculators, two things happen: it slows you down with every future topic, and it makes you feel stupid, even though you are not. So let's get these problems solved once and for all.

FRACTIONS (POSITIVES ONLY HERE)

One of the big problems with fractions is that many students do not know what a fraction is (a question the SAT loves to ask). Let's get going.

Suppose I am 6 years old. I ask you, "What is 3/7?" Oh, I am a smart 6 year old. I'll give you a start.

Suppose I have a pizza pie. . . . We divide it into 7 equal parts, and we count 3 out of the 7 parts. That is 3/7. Which is bigger, 2/7 or 3/7?

2/7 means divide a pie into 7 equal pieces and count 2 out of the 7 parts. 3 is more than 2. Soooo 3/7 is greater.

Which is bigger, 3/7 or 3/8? We know that 3/7 is. Divide the pizza into 8 parts and count out 3 slices. If you divide a pie into 8 parts, the pieces are *smaller!!!!* So 3/7 is bigger than 3/8.

We get the following rule: If the *denominators* (bottoms) are the same, the bigger the top, the bigger the fraction. If the *numerators* (tops) are the same, the bigger the bottom the *smaller* the fraction.

Now that we finally know what a fraction is, we can reduce them, multiply them, divide them, add them, and subtract them.

Reducing

Method 1: Suppose we want to reduce a fraction. We find a number that divides into the top and also the bottom.

EXAMPLE I

6/9. 3 divides into both 6 and 9. 6/3 = 2. 9/3 = 3. 6/9 = 2/3.

EXAMPLE 2

48/60. It is not necessary to find the largest number if you don't see it. You might see that 4 works. So 48/60 = 12/15. Now we see that 3 works. Soooo 48/60 = 12/15 = 4/5. (12 would have worked, but don't worry if you didn't see it. As you practice, you will see more.)

Method 2: 48/60. First write 48 and 60 as the product of prime factors:

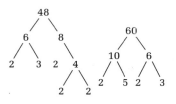

$$48 = 2 \times 2 \times 2 \times 2 \times 3 \qquad 60 = 2 \times 2 \times 3 \times 5$$

$$\frac{48}{60} = \frac{2 \times \cancel{2} \times 2 \times \cancel{2} \times \cancel{3}}{\cancel{2} \times \cancel{2} \times \cancel{3} \times 5} = \frac{4}{5}$$

It is not the shortest method, but it is necessary for two reasons. The first is in case the numerators and denominators are big. The second is the way we work with algebraic fractions.

Multiplication and Division

I believe that multiplication and division of fractions should always be taught first because they are easier and because they are needed for adding and subtracting. Because adding is taught first and lots more time is spent on it, many of you have trouble with this topic.

Rule: $\dfrac{a}{b} \times \dfrac{c}{d} = \dfrac{ac}{bd}$

Multiply the tops and multiply the bottoms.

EXAMPLE 1

$$\frac{3}{4} \times \frac{5}{7} = \frac{15}{28}$$

EXAMPLE 2

Sometimes you have to reduce before you multiply:

$$\frac{\overset{2}{\cancel{4}}}{15} \times \frac{25}{\underset{3}{\cancel{6}}} = \frac{2}{\underset{3}{\cancel{15}}} \times \frac{\overset{5}{\cancel{25}}}{3} = \frac{2}{3} \times \frac{5}{3} = \frac{10}{9}$$

Definition

Product: **Answer in multiplication.**

Factor: **If a × b = c, a and b are factors of c. If 6 × 3 = 18, then 6 and 3 are factors of 18.**

Natural numbers: **1, 2, 3, 4 . . .**

Prime: **A natural number with exactly *two, distinct* factors, itself and 1. 1 is not a prime because it has only one factor: 1 × 1 = 1. 2 is a prime because 1 × 2 = 2. 3 is prime, butttt 4 is not a prime: 4 = 1 × 4 or 2 × 2. 4 has three factors, 1, 2, and 4. It might be worthwhile to know the first eight primes: 2, 3, 5, 7, 11, 13, 17, 19.**

NOTES

1. You can't cancel the tops.

2. You would do all cancelling in one step.

NOTE

You may reduce the top and bottom of the same fraction.

EXAMPLE 3

$$\frac{4}{9} \times \frac{15}{\underset{25}{\cancel{50}}} = \frac{2}{\underset{3}{\cancel{9}}} \times \frac{\overset{5}{\cancel{15}}}{25} = \frac{2}{3} \times \frac{\overset{1}{\cancel{5}}}{\underset{5}{\cancel{25}}} = \frac{2}{3} \times \frac{1}{5} = \frac{2}{15}$$

EXAMPLE 4

$$4 \times \frac{3}{125} = \frac{4}{1} \times \frac{3}{125} = \frac{12}{125}$$

Division rule: $\dfrac{a}{b} \div \dfrac{c}{d} = \dfrac{a}{b} \times \dfrac{d}{c} = \dfrac{ad}{bc}$

Invert (flip upside down) the second fraction and multiply, canceling if necessary.

EXAMPLE 5

$$\frac{4}{7} \div \frac{5}{11} = \frac{4}{7} \times \frac{11}{5} = \frac{44}{35}$$

You might ask, "Why do you invert?" A good question! Here's an explanation:

EXAMPLE 6

$6 \div 2$. This means

$$\frac{6}{1} \div \frac{2}{1} = \frac{6}{1} \times \frac{1}{2} = \frac{3}{1} = 3$$

Dividing by 2 is the same as multiplying by 1/2.

Definiton & (ampersand): And.

Adding & Subtracting

Adding and subtracting are done the same way. So we'll concentrate on addition.

Rule: $\dfrac{a}{c} + \dfrac{b}{c} = \dfrac{a + b}{c}$ $\dfrac{3}{7} + \dfrac{2}{7} = \dfrac{5}{7}$

The problem is when the bottoms are different. We must find the least common denominator, which in reality is the *least common multiple* (LCM).

Multiple (nn): **Take all nn and multiply by a nn. Multiples of 5: 5 × 1, 5 × 2, 5 × 3, 5 × 4, 5 × 5, 5 × 6, . . . = 5, 10, 15, 20, 25, 30, . . .**

EXAMPLE 1—

Find the LCM of 2 and 3.

The LCM consists of three words. The last is *multiple.*

Multiples of 2 (evens): 2, 4, ⑥, 8, 10, ⑫, 14, 16, ⑱, 20, 22, ㉔, 26, . . .

Multiples of 3: 3, ⑥, 9, ⑫, 15, ⑱, 21, ㉔, 27, . . .

The next word is *common* (multiples): 6, 12, 18, 24, 30, . . .

The last word is *least* (common multiple): 6.

There are three techniques for adding fractions: one for small denominators, one for medium denominators, and one for large bottoms.

EXAMPLE 2—

$$\frac{3}{4} + \frac{1}{6}$$

LCD of 4 and 6, you must see, is 12.

$$\frac{3}{4} = \frac{9}{12}$$

4 into 12 is 3. 3 × 3 = 9.

$$\frac{1}{6} = \frac{2}{12}$$

6 into 12 is 2. 1 × 2 = 2.

$$\frac{3}{4} + \frac{1}{6} = \frac{9}{12} + \frac{2}{12} = \frac{11}{12}$$

Adding Medium Denominators

EXAMPLE

Add: 1/4 + 1/6 + 3/8 + 2/9. What is the LCD?

It is hard to tell, although some students will be able. What should you do?

Take multiples of the largest bottom, 9. 9, 18 (4 doesn't go into 18), 27, 36 (4 goes into; so does 6; but 8 does not), 45, 54, 63, 72. 72 works!!!!

If you must—I don't require it, but if your teacher wants, do it.

$$\frac{1}{4} = \frac{18}{72}$$

$$\frac{1}{6} = \frac{12}{72}$$

$$\frac{3}{8} = \frac{27}{72}$$

$$\frac{2}{9} = \frac{16}{72}$$

Total is $\frac{73}{72} = 1\frac{1}{72}$

Not too bad so far.

Adding Large Denominators

We need to know how to add when the LCD is large because it is the same process we need for algebraic fractions. Unfortunately, it is much messier for numbers!!!! This skill is very important!!!!

EXAMPLE

Add: $\dfrac{3}{100} + \dfrac{5}{48} + \dfrac{25}{54} + \dfrac{1}{30}$

What is the LCD of 100, 48, 54, 30. You should know. . . . I'm just kidding. Almost no one, including me, knows the answer. So here's the steps for adding the fractions:

Step 1. Factor all denominators into primes.

$100 = 2 \times 2 \times 5 \times 5$

$48 = 2 \times 2 \times 2 \times 2 \times 3$

$54 = 2 \times 3 \times 3 \times 3$

$30 = 2 \times 3 \times 5$

Step 2. The "magical" phrase. The LCD is the product of the most number of times a prime appears in any *one* denominator). 2 appears twice in 100, four times in 48, once in 54, and once in 30. LCD has four 2s. 3 appears no times in 100, once in 48, three times in 54, once in 30. The LCD has three 3s. The LCD has two 5s. LCD = $2 \times 2 \times 2 \times 2 \times 3 \times 3 \times 3 \times 5 \times 5$. Whew!!!!

Step 3. Multiply top and bottom by "what's missing." (Look at the example at the side.)

Another way to look at Example 1:

$$\frac{3}{4} + \frac{1}{6} = ?$$

$$\frac{3}{4} = \frac{3 \times 3}{4 \times 3} = \frac{9}{12}$$

$$\frac{1}{6} = \frac{1 \times 2}{6 \times 2} = \frac{2}{12}$$

$$\frac{11}{12}$$

Let's take a look at

$$\frac{25}{54} = \frac{25}{2 \times 3 \times 3 \times 3} \qquad \text{LCD} = 2 \times 2 \times 2 \times 2 \times 3 \times 3 \times 3 \times 5 \times 5$$

54 has one 2. LCD has four 2s. $2 \times 2 \times 2$ is missing. 54 has three 3s. LCD has three 3s. No 3s are missing. 54 has no 5s. LCD has two 5s. 5×5 is missing. Multiply top and bottom by what's missing.

$$\frac{25}{54} = \frac{25}{2 \times 3 \times 3 \times 3} = \frac{25 \times 2 \times 2 \times 2 \times 5 \times 5}{2 \times 3 \times 3 \times 3 \times 2 \times 2 \times 2 \times 5 \times 5}$$

$$= \frac{25 \times 2 \times 2 \times 2 \times 5 \times 5}{2 \times 2 \times 2 \times 2 \times 3 \times 3 \times 3 \times 5 \times 5}$$

Do the same for each fraction.

Unfortunately, the next step with numbers is much messier than with letters. You must multiply out all the tops and bottoms.

$$\frac{3}{100} = \frac{3}{2 \times 2 \times 5 \times 5} = \frac{3 \times 2 \times 2 \times 3 \times 3 \times 3}{2 \times 2 \times 2 \times 2 \times 3 \times 3 \times 3 \times 5 \times 5}$$

$$= \frac{324}{10,800}$$

$$\frac{5}{48} = \frac{5}{2 \times 2 \times 2 \times 2 \times 3} = \frac{5 \times 3 \times 3 \times 5 \times 5}{2 \times 2 \times 2 \times 2 \times 3 \times 3 \times 3 \times 5 \times 5}$$

$$= \frac{1,125}{10,800}$$

$$\frac{25}{54} = \frac{25}{2 \times 3 \times 3 \times 3} = \frac{25 \times 2 \times 2 \times 2 \times 5 \times 5}{2 \times 2 \times 2 \times 2 \times 3 \times 3 \times 3 \times 5 \times 5}$$

$$= \frac{5,000}{10,800}$$

$$\frac{1}{30} = \frac{1}{2 \times 3 \times 5} = \frac{1 \times 2 \times 2 \times 2 \times 3 \times 3 \times 5}{2 \times 2 \times 2 \times 2 \times 3 \times 3 \times 3 \times 5 \times 5}$$

$$= \frac{360}{10,800}$$

Last steps. Add the top and reduce. If we add we get

$$\frac{6,809}{10,800}$$

We now have to reduce it. No, I'm for real. It is not as bad as it looks. 10,800 has only three prime factors, 2, 3, and 5. 6809 is not divisible by 2 because the last digit is not even. 6809 is not divisible by 5 because the last digit is not 0 or 5.

There is a trick for 3 (same for 9): $6 + 8 + 0 + 9 = 23$. Because 23 is not divisible by 3 (or 9), neither is 6809. The SAT likes these tricks. Adding algebraic fractions is much less messy! The steps are the same. So try a few of these.

DECIMALS

Probably the easiest of the three (fractions, decimals, and percents), is decimals. The only difficulty is reading decimals. The secret is to read the part after the decimal point as if it were a whole number and then say the last place.

EXAMPLE 1

.76   76   *Hundredths*

EXAMPLE 2

.006   6   *thousandths*

EXAMPLE 3

.4567   Four thousand, five hundred sixty-seven *ten-thousandths*

Let us put one big number down.

Hundred millions	Ten millions	Millions	Hundred thousands	Ten thousands	Thousands	Hundreds	Tens	Units	Tenths	Hundredths	Thousandths	Ten thousandths	Hundred thousandths	Millionths
1	2	3 ,	4	5	6 ,	7	8	9 .	7	6	5	4	3	2

The number is read, "one hundred twenty-three million, four hundred fifty-six thousand, seven hundred eighty-nine *and* seven hundred sixty-five thousand, four hundred thirty-two millionths."

NOTE

You only say the word *and* when you get to the decimal point and no other time. If you listen to the radio or TV, hear how many times people say large numbers incorrectly by saying that word *and* when they shouldn't. It happens many, many, many times.

Adding and Subtracting Decimals

This is really easy. Line up the decimal point and add (or subtract).

EXAMPLE I

$46.7 + 1.732 + 68$

68 is a whole number.

Soooo $68 = 68$.

$$
\begin{array}{r}
46.7 \\
1.732 \\
68. \\
\hline
116.432
\end{array}
$$

EXAMPLE 2

$45.6 - 7.89$

$$
\begin{array}{r}
45.6 \\
-7.89 \\
\hline
\end{array}
=
\begin{array}{r}
45.60 \\
-7.89 \\
\hline
37.71
\end{array}
$$

Multiplying and Dividing Decimals

EXAMPLE I

$45.67 \times .932$

Steps: Multiply $4567 \times 932 = 4{,}256{,}444$

45.67 Decimal is 2 places to the left

$.932$ Decimal is 3 places to the left

Answer is $2 + 3 = 5$ places to the left

Answer is 42.56444.

But did you ever want to know why you add the places? This is a multiplication question. Well here is the reason.

The first number is hundredths. The second number is thousandths. If you multiply $1/100 \times 1/1000$, the answer is $1/100,000$, 5 places. That's all there is to it.

EXAMPLE 2

Divide 327.56 by .004.

$$.004 \overline{)327.56}$$

$$.004. \overline{)327.560.}$$

Answer is 81,890.

Did you ever wonder why you move the points in the division problem? We want to divide by a whole number.

327.56 divided by .004

$$= \frac{327.56}{.004} \times \frac{1000}{1000} = \frac{327560}{4} = 81,890$$

Multiply the bottom by 1000 (3 places to the right). If you multiply the bottom by 1000, you must multiply the top by 1000 because $1000/1000 = 1$. Only multiplying by 1 keeps the fraction the same. That's all there is here.

Decimals to Fractions and Fractions to Decimals

Decimals to fractions: "Read it and write it."

34.037. The number is read "34 and 37 thousandths." $34^{37}/_{1000}$. That's it.

Fractions to decimals: Divide the bottom into the top decimal point 4 zeros.

EXAMPLE—

$$\frac{3}{8} = \overset{.375}{8)3.0000} \qquad \frac{1}{6} = \overset{.1666}{6)1.0000} = .1\overline{6}$$

NOTE

$.\overline{3} = .33333333\dots$

$.1242424\dots = .1\overline{24}$

PERCENTS, THE BASICS

The topic that most students seem to have lots of trouble with is this one. So let me first show how to go from percent to decimals to fractions, and so on. Then we'll try to show two different methods you have never seen.

Back and Forth from and to Percents

Percent means 1/100. So 3 percent means 3/100 or .03.

Percent to decimal: Move decimal point two places to the left and drop % sign.

$$45\% = .45.\% = .45$$

$$2\% = .02.\% = .02$$

$$.45\% = .00.45\% = .0045$$

Decimal to percent: The reverse. Move decimal point two places to the right and add a % sign.

$$.32 = .32.\% = 32\%$$

$$4.5 = 4.50.\% = 450\%$$

$$73 = 73.00. = 7300\%$$

NOTE

.45% is a percent, not a decimal.

Percent to a fraction: Divide by 100% and reduce.

$$32\% = \frac{32\%}{100\%} = \frac{32}{100} = \frac{8}{25}$$

Fraction to percent: Go from fraction to decimal to percent. Divide the bottom into the top decimal point 4 zeros and then move decimal point two places to the right and add a % sign.

$$\frac{7}{8} = 8\overline{)7.0000} = .875 = .87.5\% = 87.5\%$$

Now let's learn how to do percent problems once and for all. Most people love the first new way. Some like the second.

Method 1.   Draw a pyramid and label as shown. That's all there is to it. Let's try the three basic problems.

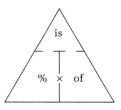

EXAMPLE 1—

What is 42% of 93?

42 is the percent. Write it as a decimal in the "percent" box. "Of" is 93. According to the chart, we multiply. .42 × 93 = 39.06. That's all!!!

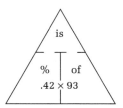

EXAMPLE 2—

12% of what is 54?

12% = .12 in the "percent" box. 54 in the "is" box. It tells you to take 54 and divide it by .12.

$$.12\overline{)54} = 12\overline{)5400}$$

Answer is 450. That's it.

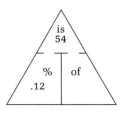

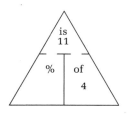

Note: The object is to get the pyramid in your head so it won't take too long.

EXAMPLE 3—

11 is what percent of 4?

11 in the "is" box. 4 in the "of" box. 11 divided by 4.

$4\overline{)11.0000} = 2.75 = 2.75.\% = 275\%$

That is all the basic problems. No more!!!!

The only problem with this method is that it doesn't explain *why*. The next method explains why. You need to know two more facts:

1. The word "of" means multiply.

2. Two fractions are equal if you cross multiply. (You probably know this now.)

$$\frac{a}{b} = \frac{c}{d} \qquad \text{if} \qquad a(d) = b(c)$$

$$\frac{5}{7} = \frac{10}{14} \qquad \text{because} \qquad 5(14) = 7(10)$$

Method 2.   a % of b is c.

$$\frac{a}{100} \times b = c$$

EXAMPLE I—

What is 42% of 93?

Let n stand for the missing number.

$$n = \frac{42}{100} \times 93 = \frac{42}{100} \times \frac{93}{1} = \frac{3906}{100} = 39.06$$

EXAMPLE 2—

12% of what is 54?

$$\frac{12}{100} \times \frac{n}{1} = \frac{54}{1} \qquad \frac{12n}{100} = \frac{54}{1} \qquad \frac{12n}{12} = \frac{5400}{12} \qquad n = 450$$

EXAMPLE 3

11 is what percent of 4?

$$\frac{11}{1} = \frac{n}{100} \times \frac{4}{1} \qquad \frac{11}{1} = \frac{4n}{100} \qquad \frac{4n}{4} = \frac{1100}{4} \qquad n = 275\%$$

I know. I sneaked in a little algebra. But you will find out that for some people algebra is easier than arithmetic. But your arithmetic must be pretty good.

Lastly, it is very useful to know some of the decimal, percentage, and fraction equivalents. It will be helpful on the SATs.

The ones you should know	You need to know these also
$25\% = .25 = \frac{1}{4}$	$12\frac{1}{2}\% = .125 = \frac{1}{8}$
$50\% = .5 \ = \frac{1}{2}$	$16\frac{2}{3}\% = .1\overline{6} \ = \frac{1}{6}$
$75\% = .75 = \frac{3}{4}$	$33\frac{1}{3}\% = .\overline{3} \ \ \ = \frac{1}{3}$
$10\% = .1 \ = \frac{1}{10}$	$37\frac{1}{2}\% = .375 = \frac{3}{8}$
$20\% = .2 \ = \frac{1}{5}$	$62\frac{1}{2}\% = .625 = \frac{5}{8}$
$30\% = .3 \ = \frac{3}{10}$	$66\frac{2}{3}\% = .\overline{6} \ \ = \frac{2}{3}$
$40\% = .4 \ = \frac{2}{5}$	$83\frac{1}{3}\% = .8\overline{3} \ \ = \frac{5}{6}$
$60\% = .6 \ = \frac{3}{5}$	$87\frac{1}{2}\% = .875 = \frac{7}{8}$
$70\% = .7 \ = \frac{7}{10}$	
$80\% = .8 \ = \frac{4}{5}$	
$90\% = .9 \ = \frac{9}{10}$	

Finally . . .

GRAPHS

You should be able to read graphs. Let us look at three kinds of questions.

EXAMPLE 1

Each symbol stands for a certain number. For example . . .

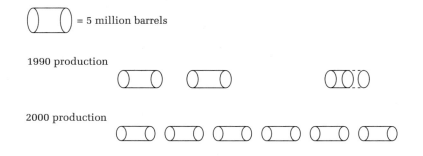

= 5 million barrels

1990 production

2000 production

How many more barrels were produced in 2000 than in 1990?

$6(5,000,000) - 2.5(5,000,000) = 30,000,000 - 12,500,000 = 17,500,000$ barrels more.

EXAMPLE 2—

The Jones family has a budget of $2000 a month. How much is paid for rent, utilities, transportation, fun, and miscellaneous?

There are 360° in a circle. Soooo

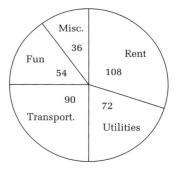

$$\frac{108}{360} = .30 \times \$2000 = \$600 \text{ on the rent}$$

$$\frac{72}{360} = .20 \times \$2000 = \$400 \text{ on the utilities}$$

$$\frac{90}{360} = .25 \times \$2000 = \$500 \text{ on transportation}$$

$$\frac{54}{360} = .15 \times \$2000 = \$300 \text{ for fun (nice)}$$

$$\frac{36}{360} = .10 \times \$2000 = \$200 \text{ for miscellaneous}$$

Notice the sum is $2000.

Let's try the reverse of Example 2.

EXAMPLE 3—

A small business has the following monthly expense account: $1000 for cultural events, $400 for sporting events, and $200 a month for miscellaneous entertainment. Draw a circle graph for this business.

First, the total budget is $1600.

$$\frac{1000}{1600} \times 360° = 225° \text{ for culture.}$$

$$\frac{400}{1600} \times 360° = 90° \text{ for sporting events.}$$

$$\frac{200}{1600} \times 360° = 45° \text{ for miscellaneous.}$$

The picture looks like this:

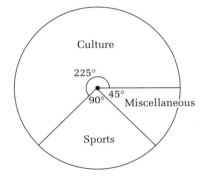

Lastly, but not leastly is . . .

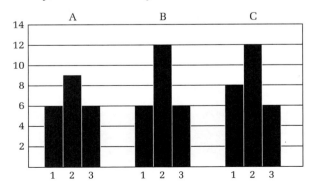

EXAMPLE 4—

Which group has a 50% increase followed by a 50% decrease?

A. From 6 to 9 is a 50% increase because (9 − 6)/6 (always over the original) is 3/6 = 1/2 = 50% buuuuut 9 to 6 is −3/9 (original—from 9 to 6) is 33⅓%. A is wrong.

B. 6 to 12 is an increase of 6. 6/6 = 1 = 100%. Already B is wrong. However, 12 to 6 is a drop of 6 and −6/12 is a drop of 50%.

C. 8 to 12 is an increase of 4/8 = 50% annnd 12 to 6 is a drop of 6 and −6/12 is a drop of 50%. The answer is C.

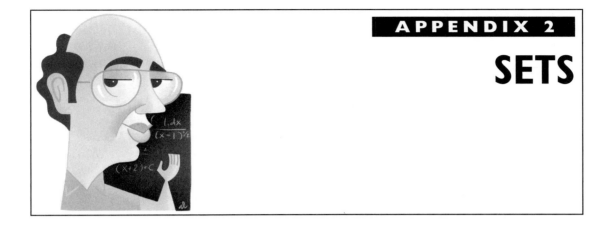

You should know something about sets, since it is a small part of algebra and the SAT.

You cannot actually define a set. We can say a *set* is a collection of "things."

The things in the set are called *elements*.

EXAMPLE 1—

{5, 6, 2} is a set containing three elements: 2, 5, 6.

Sets are indicated by braces { }.

A general set is indicated by capital letters: A, B, C, . . .

Elements are indicated by lowercase letters: a, b, c, . . .

( ) are called *parentheses* (singular: *parenthesis*).

[ ] are called *brackets*.

The *null set* (*empty set*): The set with no elements. It is indicated by { } or ϕ, the Greek letter phi.

EXAMPLE 2—

The set of all 55-foot human beings is the null set, since there aren't any—at least I don't think so.

NOTE

∈ is the Greek letter epsilon.

NOTE

Generally, a line through means "not": ≠ means "not equal."

NOTE

{0} is not, *not, NOT* the null set. It is a set with one element in it, the number zero.

NOTATION

a ∈ A is read: "A is an element of A."

EXAMPLE 3

5 ∈ {5, 6, 7}, since 5 is in the set.

NOTATION

a ∉ A is read: "a is not an element in A."

EXAMPLE 4

8 ∉ {5, 6, 7}, since 8 is not in the set.

DEFINITION

Equal sets: Set A = Set B if they have exactly the same elements.

EXAMPLE 5

A = {5, 7} and B = {7, 5}. A = B. Order does *not* matter.

EXAMPLE 6

D = {1, 1, 1, 2, 2, 1, 2} and E = {1, 2}.

D = E. Repeated elements don't count. Set D has only two elements.

DEFINITION

A ∪ B is read: "A union B" and is the set of elements in A or in B or in both.

DEFINITION

A ∩ B is read: "A intersect (ion) B" and is the set of elements in both sets.

EXAMPLE 7—

Let A = {1, 2, 3, 4, 6}, B = {2, 3, 7, 8}, and D = {7, 8, 9}.

   a. A ∪ B = {1, 2, 3, 4, 6, 7, 8}: repeated elements are
      written once.

   b. A ∩ B = {2, 3}, the common elements.

   c. A ∩ D = ϕ, since there are no elements in com-
      mon. Such sets are *disjoint.*

DEFINITION

A is a subset of B, written A ⊆ B, if every element in A
is in B.

EXAMPLE 8—

A = {1, 2} and B = {3, 2, 1}. A ⊆ B, since every element
in A is also in B.

NOTE

The whole set and the null set are subsets of any set.

EXAMPLE 9—

D = {1, 2} and E = {2, 3, 4}. D ⊄ E.

D is not a subset of E: 1 is in D but not in E.

NOTE

Also, E is not a subset of D: 3 and 4 are in E but not
in D.

If a set has n elements, there are 2^n subsets.

EXAMPLE 10—

For the set {a, b, c}, there are 2^3, or 8, subsets. Oh, let's
write them all out:

ϕ, {a}, {b}, {c}, {a, b}, {a, c}, {b, c}, {a, b, c}

Although not mentioned, there is a *universe,* U, which contains everything we are discussing.

DEFINITION

Complement of a set: the set of all elements in the universe but not in the set.

EXAMPLE 11

Let U = {1, 2, 3, 4, 5, 6}, A = {1, 3}.

A^c, read: "A complement" = {2, 4, 5, 6}.

NOTE 1

There are many notations for complements. You must check your book.

NOTE 2

Change the universe and you change the complement.

NOTE 3

Changing the universe doesn't change union, intersection, or subsets.

NOTE 4

The word *complement* is spelled with two e's. Compliment, spelled with an i, is how attractive you readers are. What a wonderful way to end the book!

That's all for now. If you are reading the arithmetic first, enjoy reading the rest of *Algebra for the Clueless.*

If you've finished reading this book and want more algebra, read *Precalc with Trig for the Clueless,* another book in the *Clueless* series.

Need to study math in preparation for the SAT? Read *SAT Math for the Clueless.*

Good luck. I hope you are really starting to enjoy math.

ACKNOWLEDGMENTS

I have many people to thank.

I thank my wife, Marlene, who makes life worth living, who is the wind under my wings.

I thank the rest of my family: children, Sheryl and Glenn, Eric and Wanda; grandchildren, Kira, Evan, Sean, Sarah, and Ethan; brother, Jerry; and parents and in-law parents, Cele and Lee, Edith and Siebeth.

I thank those at McGraw-Hill: Barbara Gilson, John Carleo, John Aliano, David Beckwith, and Maureen Walker, and Ginny Carroll of North Market Street Graphics.

I thank Martin Levine of Market Source for introducing my books to McGraw-Hill.

I thank Dr. Robert Urbanski, Bernice Rothstein, Sy Solomon, and Daryl Davis.

As usual, the last three thanks go to three terrific people: a great friend, Gary Pitkofsky; another terrific friend and fellow teacher, David Schwinger; and my cousin, Keith Robin Ellis.

ABOUT BOB MILLER ...
IN HIS OWN WORDS

After graduating from George W. Hewlett High
School, Hewlett, Long Island, New York, I received
my B.S. and M.S. in math from Polytechnic University. After my first class, which I taught as a
substitute, one student told another upon leaving,
"At least we have someone that can teach the
stuff." I was forever hooked on teaching. I have
taught at the City University of New York, West-
field State College, and Rutgers. My name is in
three editions of *Who's Who Among America's
Teachers.* No matter how bad I feel, I always feel
great when I teach. I am always delighted when
students tell me they hated math before but now
they like it and can do it. My main blessing is my
family: my fabulous wife, Marlene; my wonderful
children, Sheryl and Glenn, Eric and Wanda; and my
delicious grandchildren, Kira, Evan, Sean, Sarah, and
Ethan. My hobbies are golf, bowling, crossword puzzles, and Sudoku. Someday I hope a publisher will
allow me to publish the ultimate high school math text
and the ultimate calculus text, as this brilliant publisher did by publishing this book, so that our country
will remain number one in thinking, in math, and in
success.

To me, teaching math is always a great joy. I hope I
can give some of this joy to you.

INDEX